REVOLUTIONS and CONSTITUTIONS

By the same author:

A KINGDOM NOT OF THIS WORLD
Religion, Politics, and the English Revolution
1640-1689

REVOLUTIONS AND CONSTITUTIONS

Volume I
From Terror to Tyranny

Thomas Ripley Myers

Telos Books

Revolutions and Constitutions

Telos Books,
An imprint of the Anacyclosis Group, LLC
St. Petersburg, Florida
tel-os-books.com
telosbooks.pub

Hardcover ISBN: 979-8-9940412-7-7
Paperback ISBN: 979-8-9940412-2-2
Ebook ISBN: 979-8-9940412-3-9

Library of Congress Control Number: 2025927074

Cover Design: Author

Printed in the United States of America

DEDICATION

To Rebecca, Through Thick and Thin

CONTENTS

PREFACE

The author of Ecclesiastes states, "What has been is what will be, and what has been done is what will be done, and there is nothing new under the sun" (Eccl. 1:9). Apart from the theological significance of this profound statement, another observation arises concerning the study of history: it matters. Although time progresses in a linear direction and contains unique, monumental events (culminating in the return of Jesus), creation witnesses the repetition of patterns, movements, and interactions between humans.
-Ligonier Ministries

I did not come to the study of revolution through books. I came to it through thirty years in the United States Army, most of them in Special Forces. My profession was unconventional warfare. I trained for it, taught it, and practiced it in places where state authority had broken down, and factional violence had become ordinary. I have watched populations mobilized by ideologies that promised transformation through force, and I have seen what those promises actually deliver. Long before I had a theory of revolution, I had a memory of it.

The formal study began modestly, with a paper for a graduate class on terrorism, later published as "Terrorist to Tyrant." The question that drove it seemed simple: what happens when terrorists win? I had spent a career treating terrorism as a tactical problem, something to be found, fixed, and finished. But the question would not stay tactical. A turning point came in the course of that research, when I encountered a July 2010 lecture by Philip Mudd, then the FBI's Deputy Director for the National Security Branch, on counterterrorism strategy and al-Qaida ideology. Mudd described the core group of modern Islamic terrorists and its influence on affiliates and like-minded movements, and he was blunt about their goal: the revolutionary overthrow of the Western-dominated world order. Terrorism was merely their tactic. Reading those words, I realized that I had spent years fighting a revolution while calling it something else. That realization sent me deeper into the history of revolutionary movements and their use of terror, and I have never really come back out.

I realized that I had spent years fighting revolutionaries while calling them terrorists. That realization sent me into the history of revolutionary movements and their use of terror, and I have never really come back out.

What I found there changed how I thought about my own profession. The pattern repeated across centuries and continents: movements that seized power through terror seldom set terror aside afterward. The tactic became a governing principle. Revolutionary violence, undertaken in the name of justice or freedom, led again and again to tyranny. I came to see revolutionary terrorism not merely as a tactical problem requiring kinetic solutions but as a moral and political challenge demanding a constructive answer. Killing terrorists does not answer the grievances that produce them. Only legitimate governance does that, governance capable of addressing real complaints without requiring the destruction of every institution, of accommodating diversity without collapsing into chaos, of balancing liberty and order.

So I went looking for what good governance looks like, and that produced my previous book, *A Kingdom Not of This World: Religion, Politics, and the English Revolution, 1640-1689*. England's seventeenth-century agony, its wars between king and parliament and among religious factions, eventually yielded a constitutional settlement rather than permanent revolution. Studying that outcome reinforced a conviction that has only deepened since: constitutional governance, however slow and frustrating, produces more sustainable results than revolutionary violence ever has.

My years writing curriculum for Professional Military Education shaped this project as well. Teaching military professionals about irregular warfare and strategic competition requires a certain honesty, and the honest lessons are humbling. The most powerful military in history cannot impose political outcomes in complex cultural environments. Removing dictators does not automatically produce democracies. Intervening in other people's revolutions usually makes them worse.

I should also say plainly that I write as a Christian, and my faith shapes how I read this history. I believe there is more to life than politics, that ultimate hope lies beyond any earthly regime, and that no political order will achieve perfect justice because human nature is fallen. John Adams put the political consequence memorably: "If men were angels, no government would be necessary." Because men are not angels, both the temptation of revolution and the danger of concentrated power are permanent features of our condition. Revolutionary movements promise to perfect society from above, assuming that just institutions will produce just people. My faith teaches me the reverse. Institutions matter, but they cannot transform hearts, and sustainable change begins with the moral renewal of people and communities rather than from coercion by the state.

Why publish this now? Because the word "revolution" has come home. It is no longer something I study in archives or confront in distant valleys; it is in the mouths of my own countrymen. Politicians of both parties describe their movements as revolutionary. Activists on the left speak of burning it all down by any means necessary. Activists on the right invoke the blood of tyrant and the tree of liberty. Each side's extremism validates and empowers the other's, and the spiral of reciprocal radicalization is the most dangerous political dynamic I know.

Most of those calling for revolution, I suspect, have not counted the cost. For them it is abstract, a slogan, a posture. Having seen the real thing, I can tell them what an actual revolution brings: not social justice but death and destruction, lost friends, lost family, lost liberty. There are others, however, who understand exactly what they advocate. They believe the current order is irredeemably corrupt and that political violence is not only justified but required. These people are committed and dangerous, and history offers a sobering verdict on what they do once they hold power. They rarely renounce the violence that brought them there.

The work unfolds in three volumes. The first studies major revolutions throughout history and the forces that shaped them. The second examines counter-revolution and the role of external actors through the lens of irregular war. The third traces the development of republican government in the European tradition and closes with a theological exploration of the moral and spiritual questions that revolution and constitutional order inevitably raise.

I do not romanticize revolution, and I do not dismiss the grievances that fuel it. My aim is to understand revolution honestly, in all its complexity, and to ask what the long record of revolutions and republics can offer a deeply divided United States. The challenges are real and the polarization is genuine. But the path forward exists, if we have the wisdom to see it and the courage to walk it.

My interest is in preventing Americans from making mistakes that history shows lead to tyranny and civil war. This book is a warning not a counsel of despair. Constitutional mechanisms for reform exist. Federalism can accommodate profound differences, civic virtue can be cultivated, and moral renewal can restore communities. I would not have written these volumes if I believed otherwise.

Thomas Ripley Myers
St. Petersburg, Fl.

PROLOGUE

THE ROMAN REVOLUTION

Civil strife is as much a greater evil than a concerted war effort as war itself is worse than peace. - Herodotus

Both Cicero and Dionysius describe three types of government: Rule by the one, the few, and the many.[1] No form of political order is without both particular virtues and attendant risks. Cicero, an advocate of the rule of the few, believed rule of the many would always succumb to the power of one because "the people themselves, so dissolute and so unbridled, are always inclined to flatter a number of demagogues."[2] His analysis has stood the test of time; tyranny rather than democracy is often the result of revolutions.

The crisis known as the Roman Revolution, beginning in 133 B.C., stands as a prime example. Far from restoring the Republic, it ended in the autocracy of Caesar Augustus. The transition from Republic to Empire was, in many respects, a revolution of the people against the traditional oligarchy. Ambitious strongmen harnessed popular grievances to seize power and impose autocratic rule. Like other revolutions, it raised enduring questions about power, reform, and legitimacy, questions that still resonate today.

State Breakdown

For a state to collapse, three key elements must converge: economic crisis, elite division, and the capacity to mobilize the masses. The Roman Republic suffered from all three. The most pressing issue was agrarian: land reform.

---

[1] Cicero, *On the Republic* 1.34; all translations of this text are taken from *The Political Works of Marcus Tullius Cicero*, Francis Barham, trans. (London: Edmund Spettigue, 1841-42).; Dionysius of Halicarnassus, *Roman Antiquities* 2.3.7; all translations of this text are taken from the Loeb series, LacusCurtius site, http://penelope.uchicago.edu/Thayer/E/Roman/Texts/Dionysius_of_Halicarnassus/2A*.html.

[2] Cicero, *On the Republic* 1.34.

Roman aristocrats, especially those of the senatorial class, monopolized public lands, forming vast landed estates worked by slaves. Small farmers, the backbone of the early Republic and its military, were being displaced. Land was not just wealth but the qualification for military service and political standing.[3]

The Gracchi brothers, Tiberius and Gaius, emerged from this context.[4] Scions of Roman nobility, they chose to align with the *populares*, a political faction that claimed to represent the interests of the common people, or plebeians, by advocating reforms through popular assemblies. Their opponents, the *optimates*, were a conservative bloc of aristocrats who sought to preserve senatorial authority and the traditional structures of Roman governance.[5] The *plebes*, or plebeians, comprised the majority of Rome's free population, farmers, laborers, and artisans, whose economic stability and military service were crucial to the Republic, yet whose political rights were often curtailed. The Gracchi came from a distinguished lineage. Their father, Tiberius Sempronius Gracchus, had served with honor as consul in 177 and 163 B.C., and was celebrated for his diplomatic and military service in Spain and Sardinia. Their mother, Cornelia, was the daughter of Scipio Africanus, the hero of the Second Punic War who defeated Hannibal at the Battle of Zama. Cornelia was renowned for her intellect and virtue and played a key role in shaping her sons' values.

The Roman Senate, composed of patricians, functioned as the principal deliberative body of the Republic. While it held no formal legislative power, it exercised immense influence over public policy, foreign affairs, and the allocation of funds. In contrast, the office of the Tribune of the Plebs was created to protect the interests of the plebeians. Tribunes had the power to propose legislation and to veto the actions of magistrates and the Senate, acting as a crucial check on aristocratic dominance. The Gracchi used the office of Tribune to bypass the Senate and push for agrarian reform, seeking to redistribute public land (*ager publicus*) to dispossessed Roman citizens. Their bold confrontation with senatorial authority marked a turning point in Roman politics. In response, the Senate orchestrated their murders in 133 (Tiberius) and 121 B.C. (Gaius), ushering in an era of political violence that would become

---

[3] "Crises of the Republic," "Western Civilization", Lumen Learning, https://courses.lumenlearning.com.

[4] H.H. Schullard, *From The Gracchi To Nero: A History Of Rome from 133 BC To AD 68* (London: Methuen, 1970).

[5] D.S. Potter, *Ancient Rome: A New History,* Second ed. (N.Y., NY: Thames & Hudson, 2018), 125.

endemic.[6]

Gaius Marius, a *novus homo* (new man) from the equestrian order, rose to prominence not through aristocratic lineage but through exceptional military and political acumen. The equestrian order, or *equites*, was a social class just below the senatorial elite, composed primarily of wealthy citizens who engaged in commerce and tax farming but were barred from holding high political office unless they ascended through military or popular support.

Marius was born around 157 B.C. and later married Julia of the Julii Caesares, the paternal aunt of Julius Caesar. For a *novus homo* from Arpinum, this marriage provided aristocratic legitimacy and access to elite Roman networks. It did not make Marius a kinsman of the Gracchi in any direct sense, but it associated him with the reformist political world they had helped create. The Gracchi had exposed the connection between land distribution, military manpower, and the political rights of Rome's poorer citizens. Marius's later reforms, especially his recruitment of the propertyless and the settlement of veterans, carried those issues into the army and made them central to the final crisis of the Republic. Their legacy helped shape his populist orientation and his appeal to Rome's politically marginalized classes. His military reputation was solidified during the Jugurthine War (112–105 B.C.), when he defeated the Numidian king Jugurtha with critical assistance from his quaestor, Lucius Cornelius Sulla. That cooperative beginning, however, eventually gave way to bitter rivalry. In 88 B.C., Marius used political influence to have himself appointed commander of the Roman forces against Mithridates VI of Pontus, thereby stripping Sulla of the command. In response, Sulla defied precedent and led his legions against Rome itself, marking the first time a Roman commander had marched an army on the city. Marius returned to power after a violent coup in 87 B.C., aligning with Lucius Cornelius Cinna to purge Rome of Sullan supporters. His final months were marked by terror and executions before his death in January 86 B.C. Sulla later reclaimed Rome and imposed his own dictatorship from 82 to 79 B.C., initiating widespread proscriptions and reversing many of the popularis reforms associated with Marius and his allies.

Marius's feud with Sulla marked a critical breakdown in republican norms, as Rome's political disputes increasingly devolved into violent confrontations backed by military force. Marius's military reforms, which opened enlistment to the landless poor, created armies loyal to generals rather than Rome sowing the

---

[6] "Gracchi Brothers," "Populists of the Late Roman Republic", TheCollector, https://www.thecollector.com/gracchi-brothers-populists-roman-republic/

seeds for civil war. Coupled with the retaliatory purges and proscriptions, the Marian-Sullan era was a precursor to the later chaos under Caesar and Antony.

The Struggle

Political polarization deepened. On one side stood the *optimates*, defenders of the traditional senatorial order. On the other, the *populares*, who claimed to represent the people's interests. Yet neither faction was monolithic—many sought power above principle.

Julius Caesar's rise to power marked the most dramatic phase of the Republic's collapse into autocratic rule. Born in 100 B.C. to a patrician family of the Julian *gens*, who traced their ancestry back to the goddess Venus through Aeneas, the mythical hero of Troy considered a founder of Rome, a claim Julius would later exploit for political legitimacy. Despite his noble birth, his family was relatively impoverished and aligned with the *populares*. His aunt Julia married Gaius Marius, and Caesar would adopt the Marian cause early in his political career, earning him the enmity of Sulla.

Caesar survived Sulla's proscriptions through the intercession of influential allies, and after Sulla's death in 78 B.C., he gradually rebuilt his political standing. His early career included a brief tenure as a military officer in Asia and a celebrated episode in which he was captured by Cilician pirates, only to return and crucify his captors after securing his ransom. He steadily climbed the *cursus honorum*, Rome's traditional sequence of public offices, serving as quaestor, aedile, and eventually praetor.

In 60 B.C., Caesar formed the First Triumvirate with Pompey and Crassus, an informal alliance born of mutual political frustration. Pompey sought land for his veterans, Crassus wanted tax concessions for his clients, and Caesar desired the consulship. Cicero saw the danger posed to the Republic by such a coalition,

I warned Pompey not to form a union with Caesar, and afterwards not to sever it. By this union I saw that the power of the Senate would be broken, by its severance a civil war be provoked.[7]

Together, they bypassed senatorial opposition. Caesar's election as consul

[7] Cicero, *Letters to his Friends* 6.6; all translations of Cicero are taken from *Cicero. The Letters of Cicero; the whole extant correspondence in chronological order, in four volumes,* trans. Evelyn S. Shuckburgh (London. George Bell and Sons, 1908-1909), Perseus Digital Library.

in 59 B.C. and subsequent command in Gaul from 58 to 50 B.C. proved decisive. During this decade, he conducted the Gallic Wars, subduing various tribes and extending Rome's dominion.

As Caesar's power grew, the Triumvirate began to unravel. Crassus's death at the Battle of Carrhae in 53 B.C. and the death of Caesar's daughter Julia, who had been married to Pompey, dissolved the personal bonds holding the alliance together. With Pompey now aligned with the Senate, tensions escalated. In 49 B.C., faced with the Senate's ultimatum to disband his army, Caesar famously crossed the Rubicon River, declaring "The die is cast." This act of defiance ignited a civil war. Caesar swiftly defeated Pompey's forces at Pharsalus in 48 B.C., after which Pompey fled to Egypt and was assassinated. Caesar pursued victory, consolidating his power across the Roman world—in Egypt, Asia Minor, and North Africa, before returning to Rome. In 46 B.C., he was declared dictator, and in 44 B.C., *dictator perpetuo*, for life. His reforms included expanding the Senate, settling veterans, revising the calendar, and granting citizenship to provincials, all aimed at stabilizing the Roman state. But Caesar's concentration of power alarmed many senators, who feared the end of republican liberty. On the Ides of March (March 15), 44 B.C., a group of senators led by Brutus and Cassius assassinated him in the Senate chamber. They believed they were preserving the Republic, but instead, they plunged Rome into another cycle of civil war.

After Caesar's assassination, the Second Triumvirate, consisting of Marc Antony (Caesar's right-hand man), Octavian (Caesar's adopted son), and Lepidus (a senior commander and loyal adherent of Caesar), joined forces to avenge his death. Once they had eliminated their common enemies, Octavian quickly turned against his former allies. Lepidus was swiftly sidelined, leaving Octavian to confront Marc Antony, a seasoned and confident military commander. Although Octavian lacked Antony's battlefield experience, he relied on the strategic brilliance of his close friend, Agrippa. As importantly, Octavian proved himself a master of propaganda. Antony's alliance and romantic involvement with Cleopatra became a powerful tool for Octavian, who painted Antony as a traitor to Rome. The civil war culminated in the decisive Battle of Actium in 31 B.C., where Octavian emerged victorious. Antony and Cleopatra fled to Egypt where they committed suicide.[8]

State Reconstitution

---

[8] Ronald Syme, *The Roman Revolution*, (Oxford: Clarendon Press, 1939), 113.

The aftermath of the Battle of Actium marked the definitive end of the civil wars and the beginning of a new political order. Octavian's victory over Antony and Cleopatra removed the last significant obstacle to his supremacy. With Egypt brought under Roman control as an imperial province, its vast wealth, including the grain supply and the treasury of the Ptolemies, was now at Octavian's disposal. This consolidation of resources enabled him to secure the loyalty of the army and the urban population.

Returning to Rome in triumph, Octavian embarked on a comprehensive program of political and social reconstruction. Cognizant of the Republic's deep institutional fragility, he did not declare himself king or dictator. Instead, he presented his power as an emergency necessity that would ultimately be relinquished. In 27 B.C., he offered to restore the Republic to the Senate and people of Rome, a calculated gesture that led the Senate to grant him the honorific title "Augustus," signifying religious sanctity and reverence.

Augustus also reorganized the administration of the provinces, dividing them between senatorial and imperial domains. While the Senate retained nominal control over the more stable provinces, Augustus personally governed the frontier regions where most legions were stationed, ensuring his command over military affairs. He created a permanent standing army loyal to the emperor, introduced a praetorian guard for his personal protection, and institutionalized control over grain distribution and public works.

In civic life, Augustus emphasized the restoration of *mos maiorum*, the ancestral customs of Rome.[9] Through legal reforms, public works, and moral legislation, he sought to revitalize Roman virtues and social cohesion. He financed temples, roads, aqueducts, and monuments that linked his name to Roman renewal. His legislation on marriage and public morality reflected a desire to stabilize society after decades of chaos.

His title was not *emperor* but *princeps*, the leading or principal citizen. Perhaps most importantly, Augustus portrayed himself as the bringer of peace and stability.[10] By integrating former enemies into his regime, employing clemency where prudent, and maintaining republican facades, Augustus established a durable structure that centralized authority without provoking widespread

[9] Augustus, *Res Gestae* 1, 7, 34; all translations are taken from the Loeb series, LacusCurtius, http://penelope.uchicago.edu/Thayer/E/Roman/Texts/Augustus/Res_Gestae/1*.html.

[10] Augustus, *Res Gestae* 13.

resistance. This pragmatic blend of autocracy and tradition retained the superficial trappings of the Republic but was a transformation into an imperial monarchy.

Conclusion

The Roman Revolution began with the idealism of the Gracchi in 133 B.C. and ended with Augustus's principate in 27 B.C., transforming a decaying Republic into an imperial monarchy. Each successive phase exposed the Republic's inability to reform itself from within. Marius weaponized popular grievance against senatorial obstruction. Sulla answered populism with proscription and constitutional reaction. Caesar bypassed the constitutional order altogether, and Augustus, the most perceptive of them all, understood what his predecessors had not: that Romans would surrender the substance of liberty so long as its symbols endured. He achieved what no reformer, no warlord, and no dictator before him had managed, an enduring peace. The Roman people, exhausted by a century of civil war, chose stability over freedom.[11] Their choice shaped not only the empire that followed but set a precedent for political reinvention whose influence extends far beyond the ancient world.

This is the pattern the present book examines across the modern revolutionary landscape: revolutions undertaken in the name of liberation that produce, through their own internal logic, new forms of tyranny. What began as reform descended into factional violence, civil war, and ultimately consolidated dictatorship cloaked in the language of restoration. The sequence is not uniquely Roman. It recurs across centuries, cultures, and ideological commitments with a regularity that demands explanation.

Rome's experience reveals that republics perish not by foreign conquest but by internal dissolution. Elites divide against themselves. Popular grievances are exploited rather than resolved. Political violence becomes normalized. Constitutional constraints lose their authority. And populations, worn down by disorder, eventually choose order over liberty. These are the forces this book traces from the age of democratic revolution to the present day.

---

[11] Syme, *The Roman Revolution*, viii.

INTRODUCTION

From Terror to Tyranny

Men thinking to better their condition, are always ready to change masters, and in expectation will take up arms against any ruler; wherein they deceive themselves and find afterwards by experience that they are worse off than before.
-Machiavelli, *The Prince,*

On the night of May 28, 2020, Minneapolis burned. The third police precinct station was ablaze, evacuated hours earlier as protesters overwhelmed the perimeter. Rioters broke into surrounding businesses; Target, AutoZone, liquor stores carrying away goods while flames consumed the buildings behind them. By dawn, entire city blocks lay in ruins. The scene would repeat itself across America over the following days: Seattle, Portland, Louisville, Kenosha, Atlanta, Philadelphia. Some protesters carried signs demanding justice for George Floyd. Others threw Molotov cocktails at police stations and federal buildings. Still others proclaimed that the system itself was irredeemably corrupt, that no reform could save it, that only revolutionary transformation would suffice.

In Portland, protesters established an autonomous zone outside the federal courthouse, erecting barriers and declaring the area liberated territory. Night after night through the summer, demonstrators clashed with federal agents while politicians debated whether this was legitimate protest or insurrection. Some activists spoke openly of abolishing police departments entirely, dismantling existing structures of governance, building something new from the ruins. The language of revolution, once confined to academic journals and fringe manifestos, had entered mainstream political discourse. Major cities braced for nightly violence. Property damage exceeded two billion dollars.[12] Although it is difficult to ascertain the true number, close to 25 people died in

---

[12] Steven Silver, "How Much Damage Did the Summer 2020 Riots Really Cause?" *The National Interest*, June 10, 2022. Estimates of insured losses from the 2020 civil unrest ranged from $1 to $2 billion, making them the costliest in U.S. insurance history.

the unrest.[13] The Nation had not seen such widespread civil disorder since 1968.

Seven months later, on January 6, 2021, a different crowd gathered in Washington, D.C. they had come, they said, to defend the constitution. Many carried American flags. Some wore tactical gear and helmets. They believed that the presidential election had been stolen, that Congress was about to certify a fraudulent result, that the republic hung in the balance. When president trump finished speaking, thousands marched to the Capitol. What began as protest became a breach. Windows shattered. Doors buckled. Police officers fought in corridors while lawmakers evacuated. Rioters occupied the senate chamber, rifled through congressional offices, and chanted threats against the vice president. Others simply wandered through the halls, taking selfies, seemingly uncertain what they had come to do.

Several people died that day, and hundreds faced federal charges ranging from trespassing to seditious conspiracy. The nation watched in disbelief as the peaceful transfer of power, a cornerstone of American constitutional governance since 1797, suffered its most serious challenge in modern history.[14] Like the summer protesters before them, some January 6 participants spoke the language of revolution. They saw themselves as patriots defending liberty against tyranny. They invoked 1776. They believed they were saving the republic.

Both groups were wrong.

Between May 2020 and January 2021, Americans across the political spectrum flirted with revolutionary thinking. The left proclaimed that systemic racism required systemic dismantling. The right insisted that electoral fraud justified extraordinary measures. Each side viewed the other as an existential threat. Each accused the other of destroying democratic norms. Each, at its most extreme fringes, suggested that violence is necessary. Mainstream politicians and media personalities, rather than cooling tensions, often amplified them. Cable news showed burning buildings on loop. Social media algorithms pushed the most inflammatory content to the top. The center collapsed. Compromise became betrayal. Opponents became enemies.

This was not yet revolution. But it was closer than most Americans realized or wanted to admit.

---

[13] "What were the total number of fatalities during the George Floyd protests in 2020?" *Factually*, 15 December 2025.

[14] The first peaceful transfer of presidential power occurred on March 4, 1797, when George Washington voluntarily relinquished the presidency to John Adams. See Ron Chernow, Washington: A Life (New York: Penguin, 2010), 755–760.

The revolutionary temptation

Revolution possesses a peculiar romantic appeal, especially for those who have never experienced one. It promises a clean break from corruption and injustice, a chance to build something pure and new. Revolutionary rhetoric offers moral clarity: we are righteous, they are oppressors; we fight for justice, they defend tyranny; our cause is so urgent that ordinary rules no longer apply. For the idealist, revolution represents the triumph of principle over compromise. For the frustrated, it offers catharsis, the satisfaction of finally *doing something* rather than accepting powerlessness. For the angry, it provides permission for rage.

The appeal transcends ideology. Revolutionary fervor animated French Jacobins who sent aristocrats to the guillotine in the name of liberty, equality, and fraternity. It drove Russian Bolsheviks who promised peace, land, and bread but delivered war, collectivization, and famine. It inspired Iranian students who overthrew the shah to establish an Islamic republic, only to watch ayatollah Khomeini consolidate a theocratic dictatorship more repressive than what they had replaced. The ideologies, Enlightenment rationalism, Marxist materialism, Shi'a fundamentalism, differed but the pattern remained constant: idealism begetting violence, violence producing chaos, chaos enabling tyranny.

Contemporary Americans who invoke revolution rarely grasp what they are actually calling for. They imagine revolution as a protest movement with better outcomes, a political campaign with higher stakes, or perhaps a military coup that installs their preferred policies. Few have seriously contemplated what armed conflict in American cities would mean. Fewer still understand revolution's historical trajectory: how movements that begin with demands for justice devolve into factional warfare, how liberators become oppressors, how revolutionary governments consume their own children.

The brief experiment of the Capitol Hill Autonomous Zone (CHAZ) in Seattle offered a small but revealing glimpse into the disorder that arises when traditional governance collapses. Within days of declaring independence from city authority, the zone experienced violence, sexual assault, and ultimately multiple shootings including at least two deaths.[15] the "abolish police" movement discovered that communities still require security, dispute resolution, and protection of rights, precisely the functions that governments provide. When existing institutions disappear, they are replaced not by

[15] Mike Carter and Steve Miletich, "Shooting at Seattle's CHOP protest site kills 16-year-old boy, leaves 14-year-old seriously injured," The Seattle Times, June 29, 2020.

spontaneous harmony, but by whoever wields force most effectively.

The pattern recurs throughout revolutionary history. The Jacobins guillotined moderates and rivals as enthusiastically as they executed aristocrats.[16] Stalin purged Bolsheviks who had made the revolution alongside him.[17] Khomeini eliminated the liberals and leftists who helped overthrow the shah.[18] In each case, revolutionaries discovered that destroying existing authority creates power vacuums filled by those willing to use maximum violence. The most ruthless factions win revolutionary struggles because they alone possess the ideological certainty and moral flexibility to do whatever revolution demands.

Calls for revolution may appeal to those frustrated with the status quo, but the reality is far more harrowing than they imagine. Revolutions tend to be unpredictable and violent. Instead of ushering in social justice or democratic renewal, they often result in prolonged chaos, authoritarianism, or civil war. Friends and families are torn apart. Lives and liberties are lost. Revolution is not simply about protesting in the streets or disrupting political processes—it is about the wholesale destruction of existing order with no guarantee that what follows will be better, and substantial historical evidence that it will be worse.

The warning

This volume warns Americans against revolutionary radicalism at home. Politics is the process by which a society organizes itself, and government is the mechanism by which political decisions are implemented. In the United States, the government is intended to be "of the people, by the people, for the people."[19] But within such a system, consensus is rare. Citizens often disagree not only on policies but on the very nature and limits of government itself.

Today, progressives and conservatives are locked in an increasingly intense ideological conflict. Each side offers legitimate concerns and aspirations, but both are vulnerable to radicalization. Political radicals, on either end of the spectrum, are those who become so committed to a utopian vision that they are willing to justify extreme measures, including violence, and to dehumanize their

---

[16] Simon Schama, Citizens: A Chronicle of the French Revolution (New York: Vintage, 1989), 701–835. See also François Furet, Interpreting the French Revolution, trans. Elborg Forster (Cambridge: Cambridge University Press, 1981).

[17] Robert Conquest, The Great Terror: A Reassessment (Oxford: Oxford University Press, 2008).

[18] Ervand Abrahamian, A History of Modern Iran (Cambridge: Cambridge University Press, 2008), 181–185.

[19] Abraham Lincoln, "Gettysburg Address," November 19, 1863.

opponents.

The legitimacy of elections is essential to stable governance. When the electoral process is perceived as broken or illegitimate, some citizens may resort to more extreme measures—such as protest, resistance, or even revolution. Yet Americans possess constitutional mechanisms for reform. Federalism allows states to serve as laboratories of democracy. Separation of powers prevents any branch from dominating. The Bill of Rights protects fundamental liberties. Elections provide regular opportunities for peaceful change. These institutions have survived civil war, economic depression, and profound social transformation. We can survive current polarization, if Americans choose constitutional processes over revolutionary impulses.

Balance between the centralized state and self-government

Understanding why some revolutions produce constitutional stability while others descend into tyranny requires reframing our political categories. The conventional American political spectrum, which positions conservatives on the right and progressives on the left, suggests a linear progression or opposition. But these terms often mask more fundamental disagreements. In reality, the deeper conflict in political thought and the key to understanding revolutionary outcomes is between forces of centralization and decentralization.

The conventional labels "progressive" and "conservative" are ultimately misleading. The true fault line in political thought is between centralization and decentralization of power. The opposite of "conservative" is not "progressive," but "radical." Nor is "conservative" synonymous with "reactionary." Reactionaries long to return to a bygone era; conservatives prefer to move forward with caution and deliberation, while radicals are eager to press ahead without restraint.

Put another way the question concerns authority: where should power reside? While we may agree in principle that sovereignty rests with the people, do citizens actually possess the means to exercise that sovereignty? Excessive centralization tends to erode popular sovereignty, while excessive decentralization risks chaos and lawlessness. The enduring debate is over how much power should be vested in central authorities, how much reserved for intermediate institutions, and what decisions should be left to individuals and local communities.

Revolutionary movements are fundamentally contests over sovereignty: who holds power, how it is legitimated, and will it be centralized in the hands of a revolutionary vanguard or dispersed through constitutional limits.

Competing worldviews: riders of the apocalypse and prophets of the millennium

These differences regarding centralization and human nature are rooted in divergent worldviews. In *The Great Debate*, Yuval Levin" compared the political philosophy of Edmund Burke and Thomas Paine.[20] The French Revolution exemplifies Paine's ideas, and the American revolution Burke's. Paine, the radical progressive, saw mankind as fundamentally good but corrupted by flawed government; if virtuous people govern, society will flourish. Burke, the conservative, on the other hand viewed mankind as inherently flawed, requiring government to restrain individuals from harming one another, and, just as importantly, to limit itself, since those in power are equally prone to corruption.

Building on Yuval's work, this book identifies two broad and enduring worldviews: the "riders of the apocalypse" and the "prophets of the millennium." The apocalyptic perspective sees human nature as unchanging and history as an endless cycle of the rise and fall of peoples and civilizations, a pattern marked by recurring conflict, decline, and rebirth. This view, inherently conservative and somewhat pessimistic, regards history as a series of tragic repetitions, where peace is fleeting and struggle is inevitable.

In contrast, the millennial outlook envisions humanity and history as evolving from a primitive state toward enlightenment and harmony. This is an optimistic, even utopian, belief: that, through progress, humanity will ultimately achieve a state of perpetual peace and flourishing.

Among the riders of the apocalypse are thinkers like John "Pasha" Glubb and Arnold Toynbee. Glubb argued that all great powers follow a cyclical pattern of ascent and decline, rarely lasting more than about 250 years.[21] Toynbee described history as a succession of societies passing through phases of growth, dissolution, "times of troubles," universal states, and eventual collapse. Yet Toynbee was more hopeful than Glubb, insisting that humanity could, through reason and conscious transformation, break free from this fatal cycle.[22] For both, the study of history offers not just a chronicle of doom, but lessons, both warnings and inspirations, for those willing to heed them.

The prophets of the millennium, meanwhile, see history as a story of progress from darkness to light. For them, each generation faces unprecedented challenges, requiring new solutions; the past is less a guide than a record of obstacles overcome. Primitive social forms are destined to vanish, making way

---

[20] Yuval Levin, The Great Debate: Edmund Burke, Thomas Paine, And The Birth of Right And Left (New York: Basic Books, 2014), xiv-xv.

[21] John Bagot Glubb, The Fate of Empires and Search for Survival (Edinburgh: Blackwood, 1978), 1–12.

[22] Arnold J. Toynbee, A Study of History, abridged by D. C. Somervell (Oxford: Oxford University Press, 1947), vol. 1, 254–271.

for a new and better world. Yet this camp also has its shadows, and the path to utopia may be grim. As Marx famously argued: "the communists disdain to conceal their views and aims. They openly declare that their ends can be attained only by the forcible overthrow of all existing social conditions."[23]

Ultimately, the prophets of the millennium hope for a world where exploitation is abolished and equality prevails, believing that, at his core, man is good and, with the right government, can achieve perfection. The riders of the apocalypse, by contrast, remain skeptical: they see history as a ceaseless cycle of conflict, rooted in humanity's flawed nature, and believe the best governments can do is delay self-destruction.

Despite their differences, both camps desire a society where people can live good and meaningful lives. Yet both have produced radical centralizers, Hitler and Marx among them, who believed that only violent revolution could usher in their respective utopias. The lesson is sobering whether apocalyptic or millennial, when ideals are pursued without restraint, the quest for paradise can lead, paradoxically, to catastrophe.

This pattern recurs throughout the revolutions examined in this book. Movements begin with genuine grievances and visionary goals. They promise liberation from tyranny, justice for the oppressed, a new order built on rational or divine principles. But revolutionary success requires centralizing power to overcome resistance. What begins as liberation becomes consolidation, consolidation becomes purges, purges become terror, and terror becomes the new tyranny, often more oppressive than the regime it replaced.

From Cosmos to Constitution: The Classical Roots of Revolution and Republican Government

The word *revolution* carries a history that most who use it never consider. Its Latin root, *revolvere*, means to roll back or turn around, a term applied first to the circular motion of heavenly bodies. During the Renaissance, Nicolaus Copernicus gave the word its most famous astronomical expression in *De Revolutionibus Orbium Coelestium* (1543), describing the ordered, repeating orbits of planets around the sun. For Copernicus, revolution implied no novelty, no rupture, no irreversible change. It meant, quite precisely, a return. When early modern political thinkers needed a word for the cyclical rise and fall of governments that the ancients had long described, they reached for

[23] Karl Marx and Friedrich Engels, The Communist Manifesto (London: 1848; repr. New York: Penguin Classics, 2002), 258.

Copernicus's term and applied it to Polybius's theory of *anacyclosis*. The modern political concept of revolution was born at the intersection of Renaissance astronomy and classical political theory.

Plato and Aristotle: The Foundational Tension

The tension between centralized and dispersed authority finds its earliest systematic expression in ancient Greek philosophy, and it has never been resolved. Plato's *Republic* proposes an ideal state governed by philosopher-kings, an elite class trained from youth to rule with wisdom and virtue. Rex Martin observes that a literal translation of the title would be "the arrangement of the parts that compose the polis."[24] Plato arranged those parts into three classes: the philosopher-rulers (guardians), the warriors (auxiliaries), and the productive masses (providers).[25] Power flows from the top down. Plato conceived of freedom not as political participation but as a function of knowledge: only those who had escaped the cave of ignorance and apprehended truth were truly free, and it was their duty to govern those who remained in darkness.[26] Every revolutionary ideology claiming special insight into historical laws or divine will, from the Jacobins to the Bolsheviks to contemporary technocratic elites, echoes Plato's fundamental premise: that concentrated power in the hands of the enlightened produces better outcomes than distributed power among the ignorant.

Aristotle rejected this premise. In the *Politics*, he argued that the state exists to serve individuals, not the reverse. Absolute kingship seemed to Aristotle the most effective form of government in theory but carried within itself the seeds of its own corruption, tending to become hereditary and then tyrannical. He wrote that "the individual may be misled by his passions, and no single man can attend to all the duties of government"; therefore, kingship must be moderated.[27] Aristotle categorized government into three proper forms, each

---

[24] Rex Martin, "The Ideal State in Plato's 'Republic,'" *History of Political Thought* 2, no. 1 (1981): 1.

[25] Debra Nails, "Plato's 'Republic' In Its Athenian Context," *History of Political Thought* 33, no. 1 (2012): 1.

[26] R. F. Stalley, "Plato's Doctrine of Freedom," *Proceedings of the Aristotelian Society* 98 (1998): 157–58.

[27] Aristotle, *Politics* 3.14–16; all translations taken from *The Basic Works of Aristotle*, ed.

with a corresponding corruption: kingship degenerates into tyranny, aristocracy into oligarchy, and constitutional government (polity) into democracy, which for Aristotle was functionally equivalent to mob rule.[28] He opposed oligarchy and democracy equally because both represent extreme positions, the one serving the few, the other the unruly many.

The corrective, for Aristotle, was polity: a constitutional government grounded in a broad and moderate middle class. He praised the Laconian (Spartan) constitution as the best realized example, with its dual monarchy, council of elders, and popular assembly functioning as mutual checks.[29] He further specified that an effective constitution requires three institutional elements: a deliberative assembly, an executive authority, and courts of law.[30] As Sophie Smith notes, Aristotle believed that while the few might rule to their own benefit, the many collectively desire to direct the community for the common good, and the diverse practical wisdom of many citizens would outperform the self-seeking theories of isolated elites.[31]

The two philosophers also differed on the nature of the soul, a difference that carried political consequences. Plato held that the soul is our rational self, separable from and superior to the body, providing a link to a higher reality above the temporal world.[32] Aristotle held that the soul is the essence of the person and is inseparable from the body.[33] Plato's God is the supreme source of goodness who desires perfection in all things; Aristotle's God is the unmoved mover, the ultimate cause, who neither governs creation directly nor invites imitation through political perfectionism.[34] From these metaphysical commitments flow their respective politics. Plato's ideal state is an earthly

---

Richard McKeon and C. D. C. Reeve (New York: The Modern Library, 2001).

[28] Ibid., 4.2.

[29] Aristotle, *Politics* 4.9.

[30] Ibid., 4.14–16.

[31] Sophie Smith, "Democracy and the Body Politic from Aristotle to Hobbes," *Political Theory* 46, no. 2 (2018): 175, 180.

[32] Arthur Herman, *The Cave and the Light* (New York: Random House, 2014), loc. 446, 462.

[33] Aristotle, *De Anima* 2.413; all translations taken from *The Basic Works of Aristotle*, ed. Richard McKeon and C. D. C. Reeve (New York: The Modern Library, 2001).

[34] Herman, *The Cave and the Light*, loc. 979, 1096.

approximation of eternal order, governed by those who can perceive it. Aristotle's state is a practical arrangement for human flourishing, best managed by those with experience of actual conditions.

Polybius and Anacyclosis

Polybius was an aristocratic Greek who was deported to Rome after the Romans defeated the Macedonian army at Pydna in 168 BC. In Rome, he entered the household of Scipio Aemilianus, accompanied him on campaigns, and was present at the destruction of Carthage.[35] From this vantage point he wrote his *Histories*, seeking to explain how Rome had risen to dominate the Mediterranean world in the space of a single generation.

His answer centered on Rome's constitution, and the theoretical framework he used to illuminate it was *anacyclosis*: the cyclical degeneration of governmental forms that Aristotle had mapped but Polybius systematized.[36] The sequence, as Polybius derived it from Aristotle, runs through three proper forms and three corresponding corruptions: Monarchy degenerates into Tyranny, Aristocracy into Oligarchy, and Democracy into Ochlocracy (mob rule), after which the cycle begins again.[37] No simple form of government, however well established, can resist indefinitely the corrupting tendencies Polybius identified in human nature. The very virtues that produce good government in one generation create the conditions for its corruption in the next.

---

[35] T. Rood, "Polybius," in *Narrators, Narratees, and Narratives in Ancient Greek Literature*, ed. Irene De Jong, René Nünlist, and Angus Bowie (Leiden: Brill, 2004), 163–64.

[36] Stephan Podes, "Polybius and His Theory of 'Anacyclosis': Problems of Not Just Ancient Political Theory," *History of Political Thought* 12, no. 4 (1991): 577, http://www.jstor.org/stable/26213908.

[37] Ibid.; Aristotle, *Politics* 4.2; Polybius, *Histories* 6.4, Perseus Digital Collection, https://www.perseus.tufts.edu/hopper/text?doc=Perseus:text:1999.01.0234:book=6:chapter=4.

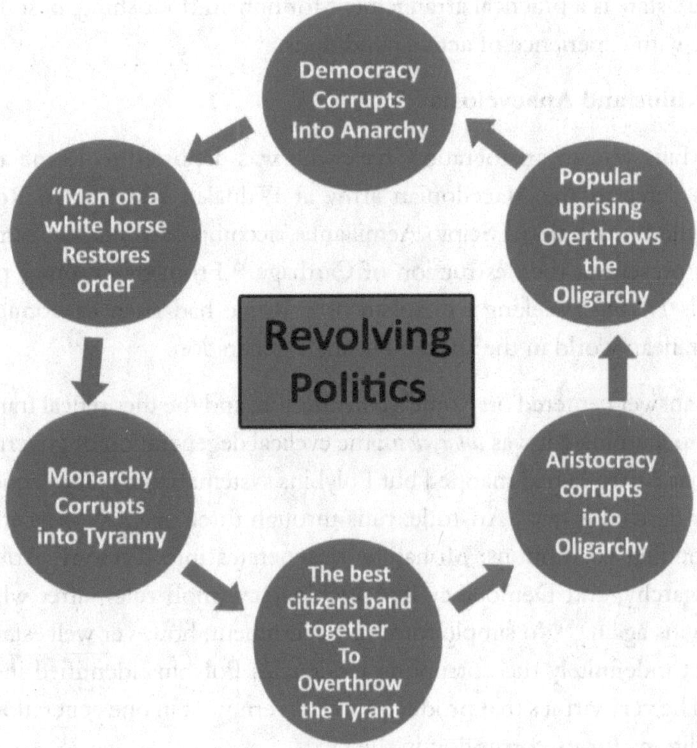

Rome, Polybius argued, had escaped this cycle by constructing a mixed constitution that balanced all three elements simultaneously. The consuls supplied the monarchic principle, the Senate the aristocratic, and the popular assemblies the democratic. Each element checked the others, and no single faction could monopolize authority long enough to corrupt it.[38] This was the insight Machiavelli would later recover, and the American founders would institutionalize.

Cicero and the Law of the Republic

Rome's political theorists did not merely describe their constitution; they argued for its moral foundations. Cicero identified the rule of law as the defining

---

[38] Polybius, *Histories* 6.4.

element in the continuity of the *res publica*.[39] Like Aristotle, he described three proper forms of government: monarchal rule by one, aristocratic rule by the best, and democratic rule by the people, each exercising a corresponding form of authority. In the Roman system, as Jill Harries explains, *potestas* lay with the magistrates, *auctoritas* with the Senate, and *libertas* with the people. These forms of control must coexist in the proper ratio to maintain stability.[40] Cicero's own preference, as Harries notes, was for the exercise of *auctoritas* by an enlightened elite, a residual Platonic instinct in an otherwise Aristotelian framework.[41]

But whatever the form of government, Cicero insisted that the vital principles were the consent of the governed and adherence to law in the service of the common good.[42] Neither principle alone suffices: consent without law degenerates into mob rule, and law without consent becomes tyranny. The term *res publica* itself, which modern usage has narrowed to mean a non-monarchical form of government, originally carried no such restriction. As James Hankins argues, in antiquity *res publica* meant the public good or common welfare, and even a monarchy could qualify as a republic if the monarch ruled for the benefit of all his subjects.[43] The narrowing of the term to its modern sense, synonymous with popular sovereignty and opposed to monarchy, was an early modern transformation, not an ancient one.

Machiavelli and the Recovery of the Cycle

---

[39] Jill Harries, "Servius, Cicero and the Res Publica of Justinian," in *Cicero's Law: Rethinking Roman Law of the Late Republic*, ed. Paul J. Du Plessis (Edinburgh: Edinburgh University Press, 2016), 123.

[40] Ibid., 137.

[41] Ibid., 138.

[42] Ibid., 134; Cicero, *On the Republic* 1.34; all translations taken from *The Political Works of Marcus Tullius Cicero*, trans. Francis Barham (London: Edmund Spettigue, 1841–42).

[43] James Hankins, "Exclusivist Republicanism and the Non-Monarchical Republic," *Political Theory* 38, no. 4 (2010): 452–82, http://www.jstor.org/stable/25704829.

The Renaissance revived classical political theory and pressed it into service for new political realities. In the *Discourses on Livy*, Niccolò Machiavelli describes the cycle of governmental forms in terms that closely follow Polybius.[44] Democratic rule of the people is gradually corrupted into anarchy. A strong citizen imposes order and institutes a monarchy. Monarchy becomes corrupt and tyrannical. The prominent citizens overthrow the tyrant and establish an aristocracy. The aristocracy is corrupted into an oligarchy. The people rise and restore democracy. And then the cycle begins again.

Machiavelli's conclusion was identical to Polybius's: the best form of government incorporates elements of each, so that "one keeps watch over the other," and this combined form is precisely what a republic achieves. Machiavelli grounded this argument not in abstract philosophy but in the concrete historical record of Rome, whose stability and power he attributed to its mixed constitution and the productive tension between the Senate and the popular assemblies.

Grotius, Natural Rights, and the Social Contract

The recovery of Aristotelian political philosophy in the late medieval and early modern period produced, through a circuitous route, the doctrine of natural rights. Although the Catholic Scholastics formally aligned themselves with Aristotle, the Church in practice operated on thoroughly Platonic assumptions of top-down authority. With the Protestant challenge to papal power, monarchs began asserting divine sanction for their own authority, and thinkers on all sides were forced to examine the foundations of political legitimacy more carefully.

Hugo Grotius carried Aristotle's premise, that the state exists to serve individuals rather than the reverse, to its logical conclusion: rights are something possessed by individuals rather than conferred by the state.[45] These natural rights could be restricted by freely agreed-upon positive law, but laws imposed by the state without the consent of the governed lack any foundation in natural

---

[44] Niccolò Machiavelli, *Discourses on Livy*, trans. Julia and Peter Bondanella (New York: Oxford University Press, 1997), 22–26.

[45] Knud Haakonssen, "Hugo Grotius and the History of Political Thought," *Political Theory* 13, no. 2 (1985): 240.

law and are therefore illegitimate.[46] Grotius was not a republican in the sense of popular sovereignty. He was a republican in the classical sense Hankins identifies: a self-governing commonwealth is not defined by its form of government but by its commitment to the rule of law and the good of the people.[47] The prince in Grotius's system functions as an intermediary rather than a sovereign in the absolute sense; what matters is that the people remain in control of the fundamental law.[48] His experience in the fractious politics of the Dutch Republic, and his subsequent exile, may have confirmed his view that sovereignty was most safely situated at the provincial level, where individual communities could cooperate or withhold cooperation at the national level.[49]

Having witnessed the chaos of the Thirty Years' War, however, Grotius also concluded that the state must be granted sufficient authority to maintain order, and that individual rights must sometimes yield to collective stability.[50] While citizens surrender some natural rights to a sovereign for the common good, they retain the right to resist sovereign authority when a ruler violates natural law.[51] The agreement to surrender rights to a sovereign in exchange for the protection of the remainder is what later thinkers would formalize as the social contract.

Hobbes, Selden, and Locke: Three Responses to Crisis

---

[46] Benjamin Straumann, "'Ancient Caesarian Lawyers' in a State of Nature: Roman Tradition and Natural Rights in Hugo Grotius's 'De Iure Praedae,'" *Political Theory* 34, no. 3 (2006): 344–45.

[47] Hankins, "Exclusivist Republicanism," 452.

[48] Daniel Lee, "Popular Liberty, Princely Government, and the Roman Law in Hugo Grotius's 'De Jure Belli Ac Pacis,'" *Journal of the History of Ideas* 72, no. 3 (2011): 391.

[49] Jan Waszink, "The Low Countries: Constitution, Nationhood and Character According to Hugo Grotius," in *The Roots of Nationalism: National Identity Formation in Early Modern Europe, 1600–1815*, ed. Lotte Jensen (Amsterdam: Amsterdam University Press, 2016), 148.

[50] Girard, *Hugo Grotius: Natural Law and Social Contract Theory*, 44–45.

[51] Ibid., 48.

The English civil war and its aftermath produced three distinct lines of political thought, each drawing on natural law theory and each representing, very loosely, one of Aristotle's three constitutional forms.

Thomas Hobbes stands in for autocracy. His foundational premise was that self-preservation is the basis of natural law, and that in the absence of a powerful ruler, human beings are reduced to a state of nature in which every man is at war with every other.[52] The sovereign's authority, however obtained, is therefore justified by the alternative, and Hobbes urged acceptance of Cromwell's protectorate on precisely these grounds. Citizens voluntarily abdicate individual rights to government in exchange for security, and the resulting Leviathan, whatever its form, is legitimate as long as it maintains order.

John Selden drew on English common law and Jewish thought to argue that humans do not derive rights as isolated individuals in a state of nature but are born into families, and it is within the family that liberty is first secured and transmitted.[53] Selden therefore advocated grounding government on traditional arrangements and inherited institutions. His vision is consistent with what later constitutional theorists called *aeque principaliter*, the preservation of regional and customary differences under a common sovereign authority, making him a defender of constitutional monarchy and the timocratic principle of rule by the many within established law.[54]

John Locke imagined a state of nature like Hobbes's in that humans join together for self-interested reasons, but unlike Hobbes he described it as a state of cooperation rather than war. Locke's government exists to protect the rights

---

[52] Straumann, "'Ancient Caesarian Lawyers,'" 337; Melvyn Bragg, "The Social Contract," discussion with Melissa Lane, Susan James, and Karen O'Brien, *In Our Time*, BBC, February 7, 2008, audio, 45:00.

[53] Isaac Herzog, "John Selden and Jewish Law," *Journal of Comparative Legislation and International Law* 13, no. 4 (1931): 236, http://www.jstor.org/stable/753376; Paul Christianson, "Young John Selden and the Ancient Constitution, ca. 1610–18," *Proceedings of the American Philosophical Society* 128, no. 4 (1984): 272; Abraham Berkowitz, "John Selden and the Biblical Origins of the Modern International Political System," *Jewish Political Studies Review* 6, no. 1/2 (1994): 29.

[54] J. H. Elliott, "A Europe of Composite Monarchies," *Past and Present*, no. 137 (1992): 52–53; Aristotle, *Nicomachean Ethics* 8.10.

of individuals who have freely chosen to form a community for mutual benefit.[55] Crucially, a ruler who fails to protect those rights is not merely bad but illegitimate; he is in effect at war with his own subjects. "Where the government is dissolved," Locke wrote, "the people are at liberty to provide for themselves, by erecting a new legislative."[56] Like Aristotle, Locke believed a virtuous citizenry was the best guarantor of both liberty and order.[57] His preference for propertied landowners as the primary bearers of political rights places him in the Aristotelian tradition of aristocracy in the sense of rule by the qualified and responsible, though the American colonists who followed him extended his principles further toward genuine popular sovereignty.[58]

Burke and Kant: The Great Divergence

By the late eighteenth century these threads had hardened into two opposing visions of political order, brought into violent confrontation by the French Revolution.

Immanuel Kant saw in revolutionary France the construction of a constitution on the basis of pure reason.[59] Where earlier theorists grounded political obligation in natural law derived from experience, Kant derived it from rational autonomy alone. He defined freedom as the absence of external governance and argued that moral law, to be binding without coercion, must

---

[55] Ethan Alexander-Davey, "Constitutional Self-Government and Nationalism: Hobbes, Locke, and George Lawson," *History of Political Thought* 35, no. 3 (2014): 463.

[56] Herman, *The Cave and the Light*, loc. 6781.

[57] John Locke, *Second Treatise of Civil Government* (1690), §220, in *Two Treatises of Government*, ed. Mark Goldie (London: Everyman, 1993), 222.

[58] David Armitage, "John Locke, Carolina, and the 'Two Treatises of Government,'" *Political Theory* 32, no. 5 (2004): 602–27, http://www.jstor.org/stable/4148117; David McNally, "Locke, Levellers, and Liberty: Property and Democracy in the Thought of the First Whigs," *History of Political Thought* 10, no. 1 (1989): 23; Ellen Meiksins Wood, "Locke Against Democracy: Consent, Representation, and Suffrage in the 'Two Treatises,'" *History of Political Thought* 13, no. 4 (1992): 657–89, http://www.jstor.org/stable/26214164.

[59] Christopher J. Insole, "Two Conceptions of Liberalism: Theology, Creation, and Politics in the Thought of Immanuel Kant and Edmund Burke," *The Journal of Religious Ethics* 36, no. 3 (2008): 448.

arise from within the individual through reason rather than from tradition, Scripture, or custom.[60] This made Kant's political theory universal: because human reason is common to all persons, the principles derived from it apply to all mankind without exception. Forming a legitimate government requires unifying "the will of one person with the will of another under a universal law of freedom," and this unified general will, a term Kant consciously borrowed from Rousseau, is the foundation of the social contract.[61] Both Kant and Rousseau shared, in this respect, a fundamentally Platonic aspiration: the perfection of society through the governance of enlightened reason.

Edmund Burke read this aspiration as catastrophic. He had dubbed Rousseau "the insane Socrates" of the French Revolution and, even before the Reign of Terror, identified in revolutionary France the inevitable working-out of the Platonic temptation: the attempt to perfect society through reason alone while ignoring what Aristotle and historical experience had taught about human nature.[62] Burke's political theory began not with the autonomous individual but with the family, the primary "little platoon" to which everyone is or should be attached and "the first principle, the germ as it were, of public affections."[63] Inherited traditions, customs, and laws are transmitted from parent to child across generations; they encode the accumulated practical wisdom of innumerable predecessors and should not be discarded because a single generation believes it has reasoned its way to something better.[64] Burke urged his countrymen to stand on the firm ground of the British constitution, by which he meant not merely a legal text but the entire corporate society represented in its local institutions and ultimately in Parliament.[65] He declared

---

60 Ibid., 459–60.

61 David Lay Williams, "Ideas and Actuality in the Social Contract: Kant and Rousseau," *History of Political Thought* 28, no. 3 (2007): 475; Insole, "Two Conceptions of Liberalism," 461.

62 Herman, *The Cave and the Light*, loc. 7645.

63 Eileen M. Hunt, "The Family as Cave, Platoon, and Prison: The Three Stages of Wollstonecraft's Philosophy of the Family," *The Review of Politics* 64, no. 1 (2002): 83–84.

64 J. Roy, "'Polis' and 'Oikos' in Classical Athens," *Greece and Rome* 46, no. 1 (1999): 1.

65 Randall B. Ripley, "Adams, Burke, and Eighteenth-Century Conservatism," *Political Science Quarterly* 80, no. 2 (1965): 228.

the French Revolution "liberty without wisdom, the greatest of all possible evils, folly, vice, and madness, without tuition or restraint."[66] If Kant's thought begins with individual autonomy, Burke's is drawn from Aristotelian empiricism and historical experience.[67] For Kant, there is only the individual and the state. Burke would ask: in a Platonic republic, who holds the guardians accountable?[68]

French Anacyclosis and the Meaning of Revolution

The French Revolution enacted Polybius's theory with remarkable precision. The Bourbon monarchy, ineffective and financially ruined, gave way first to aristocratic deliberation in the National Assembly and then to the democratic fury of the Parisian mob and the Jacobin clubs. Robespierre imposed order through the Terror, exercising the monarchic principle through revolutionary virtue rather than hereditary right. After the fall of the Jacobins, the Directory attempted to restore oligarchic stability, but popular discontent threatened anarchy once more. Napoleon finally ended the cycle by providing the strong executive that Hobbes had always argued was the indispensable alternative to chaos. In the end, it was neither Burke nor Kant nor Locke but Hobbes whose Leviathan prevailed in enlightened France.[69]

It was in this context that the Copernican term completed its migration from astronomy to politics. The word *revolution*, as Copernicus had used it in 1543, described the ordered return of a planet to its starting point: a cycle completed, not a world remade. When English constitutionalists described the events of 1688 as the Glorious Revolution, they meant precisely this: a return to the ancient liberties of the English constitution, a cycle restored rather than broken. Edmund Burke understood the American Revolution in the same terms. It was the French who broke the cycle in their own minds, adopting what Andrew Arato calls the telos of a *novus ordo seclorum*, a completely new order of the ages, replacing the cyclical time of Polybius and Copernicus with the linear

---

[66] Insole, "Two Conceptions of Liberalism," 448.

[67] Ibid., 447, 475.

[68] Melvyn Bragg, "Edmund Burke," discussion with Richard Bourke, John Keane, and Karen O'Brien, *In Our Time*, BBC, July 3, 2010, audio, 9:25, https://www.bbc.co.uk/programmes/b00sjqyn.

[69] Podes, "Polybius and His Theory of 'Anacyclosis,'" 577.

time of perfectionist ideology.[70] When a revolution believes it is completing a circle, it tends to preserve. When it believes it is inaugurating an eternal new order, it tends to destroy. The word itself, in its passage from Copernicus to Robespierre, enacts the very transformation this volume sets out to explain.

The American Synthesis

The American founders had studied the entire tradition surveyed above, and their constitutional settlement represents its most successful practical application. They distinguished sharply between a republic and a democracy. Pure democracy, in which the majority rules directly, risks the Aristotelian collapse into mob rule. Madison warned in *Federalist* No. 10 that democracies "have ever been spectacles of turbulence and contention; have ever been found incompatible with personal security or the rights of property."[71] A republic, by contrast, operates through representation and constitutional limits; elected officials deliberate on behalf of constituents rather than executing the majority's immediate will, and structural safeguards, bicameral legislature, separation of powers, federalism, and the Bill of Rights, prevent any faction from monopolizing authority.

The founders drew simultaneously on Cicero's insistence that legitimate government requires both consent and law, on Polybius's demonstration that a mixed constitution can interrupt the cycle of *anacyclosis*, on Grotius's natural rights theory, and on Locke's argument that a government forfeiting the trust of its people forfeits its legitimacy. They also drew on Burke's instinct for constitutional inheritance: the American republic was not constructed from first principles in the French manner but built upon the accumulated legal tradition of English constitutionalism. The *res publica* the founders established was, in Hankins's classical sense, a government for the common good, and in the modern sense a constitutional republic grounded in popular sovereignty and dispersed power. George Washington's voluntary relinquishment of command and then of the presidency embodied the Polybian insight: the cycle is broken

---

[70] Andrew Arato, "Revolution," *Reset Dialogues on Civilizations Intercultural Lexicon* (Milan: Reset DOC, 2012), https://www.resetdoc.org/intercultural-lexicon/revolution; Hannah Arendt, *On Revolution* (New York: Viking, 1963).

[71] James Madison, *Federalist* No. 10, in *The Federalist Papers*, ed. Clinton Rossiter (New York: Mentor, 1961), 81.

not by institutional design alone but by the virtue of citizens who refuse the temptation of concentrated power.

The *res publica* is not merely a form of government. It is a claim about the rightful relationship between authority and the people, a claim first made in the Athenian assembly, refined in the Roman Senate, recovered in the Renaissance, tested in the Reformation, theorized by Grotius and Locke, defended by Burke, and institutionalized, however imperfectly, in Philadelphia in 1787. The chapters that follow trace how that claim was made, contested, and transformed across the centuries of crisis and innovation that connect ancient Athens to the modern world.

Legitimacy: the foundation of political order

Whether power centralizes or disperses, all governments face the same fundamental question: what makes authority legitimate? This question becomes especially acute during revolutions, when existing legitimacy collapses and competing factions claim the right to rule.

Cicero insisted that legitimate government rests on two pillars: the consent of the governed and fidelity to law for the common good.[72] Neither alone suffices. Government, by consent but without law, degenerates into mob rule. Government by law but without consent becomes tyranny. Legitimate authority requires both.

Revolutionary movements contest the legitimacy of existing authority exposing corruption, injustice, or failure to serve the common good. They construct alternative narratives offering justice, dignity, or national renewal. They mobilize mass support to demonstrate popular consent for revolutionary transformation. They seize state institutions to implement their vision.

But seizing power and exercising it legitimately proves very different challenges. Revolutionary movements that succeed through terror and violence often consolidate power through the same means, foreclosing the democratic processes they claimed to champion. The radical vanguard that seizes power is rarely willing to give it up afterward. George Washington's voluntary relinquishment of power, grounded in his moral character and commitment to constitutional principles, helped America to escape the tyranny that consumed France, where revolutionary leaders like Robespierre conceived of legitimacy in

---

[72] Cicero, On the Commonwealth (De re publica), trans. James E. G. Zetzel (Cambridge: Cambridge University Press, 1999), Book I, chs. 25–34.

terms of revolutionary virtue enforced through terror.

Gulags and guillotines: the road to Robespierre is paved with good intentions

The prophets of the millennium genuinely believe what they are doing is morally justified and their reasoning to be unimpeachable. They see themselves as champions of reason or justice, while those who disagree are dismissed as either ignorant, requiring re-education or immoral, and thus deserving of exclusion and punishment. This mindset leads to a dangerous polarization: progressives may label conservatives as "far-right," "authoritarian," "fascists," or "nazis." Such labels do more than stigmatize; they imply that preemptive hostility is not only justified, but morally necessary.

Yet this is precisely the problem. *We, the people*, are individuals, not statistics or abstract categories. Real people have names: Joe, Jane, Anna. They are our friends, neighbors, and colleagues, people with whom we share coffee, conversation, and community. When we reduce others to mere statistics or faceless groups, it becomes all too easy to dehumanize and demonize them. And if we label those who disagree as evil, how can we ever hope to reconcile or find common ground?

Friedrich Hayek, in *The Road To Serfdom*, lamented this tragic tendency:

> *It is one of the saddest spectacles of our time to see a great democratic movement support a policy which must lead to the destruction of democracy and which meanwhile can benefit only a minority of the masses who support it.*[73]

The true threat to liberty and justice is not found in capitalism or socialism per se, but in the unchecked centralization of power, whether in the hands of the state, the market, or an ideological elite. Only by recognizing the humanity of our opponents, and by resisting the temptation to reduce people to labels or categories, can we preserve both democracy and decency.

Revolutionary movements across history have fallen into this trap. The Jacobins believed they represented the general will and therefore possessed authority to execute anyone who opposed them. The Bolsheviks believed historical laws justified eliminating entire classes of people. The Khomeinists believed divine authority sanctioned purging anyone who contaminated the Islamic Republic. In each case, good intentions and revolutionary fervor produced gulags and guillotines.

---

[73] Friedrich A. Hayek, The Road to Serfdom (Chicago: University of Chicago Press, 1944), 171.

This volume contends that the dominant framework for understanding political contest, a horizontal line with communists anchored on the left and fascists on the right, is not merely imprecise but actively misleading. It obscures the most important truth about political extremism: that the two ends do not stand in opposition to each other but converge on the same destination.

A circular model serves the evidence better. Place constitutional republicanism and liberal democracy at the base of the circle, where power is dispersed, law constrains ambition, and the consent of the governed provides the foundation of legitimate authority. Moving along either arc of the circle, whether leftward through pure democracy to socialism through anarchy toward communism, or rightward through bureaucracy to nationalism through oligarchy toward fascism, the traveler ascends toward the same point at the top: Totalitarian Tyranny. The ideological vocabularies differ. The uniforms differ. The enemies designated for liquidation differ. But the destination, the concentration of power in the hands of a vanguard willing to use unlimited violence in pursuit of a perfect social order, is identical.

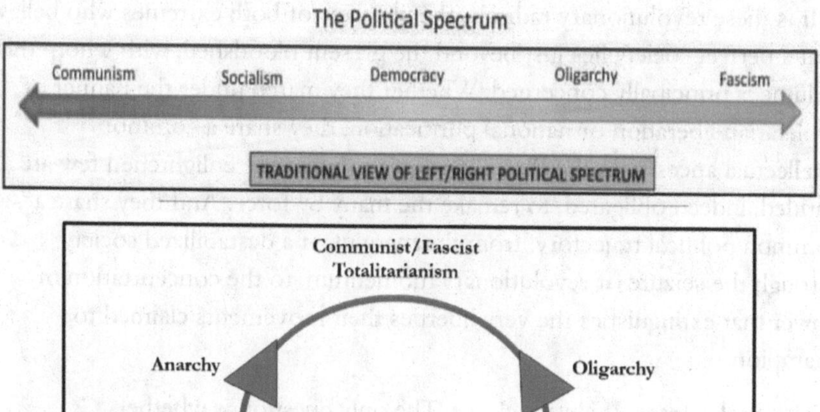

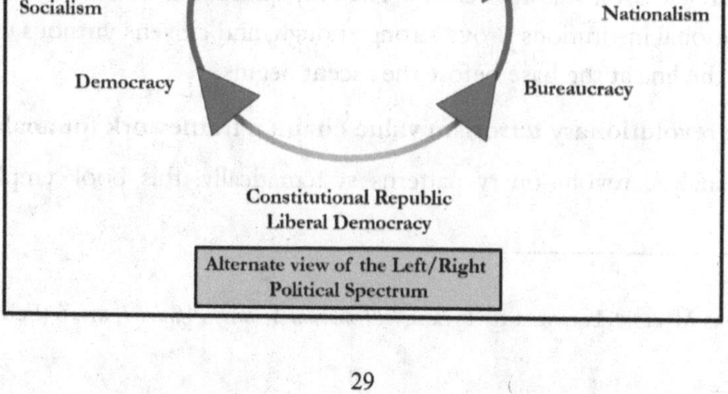

This is Polybius' *anacyclosis*. The classical theorists understood what the modern left-right schema conceals: that the real axis of political danger is between constitutional order and its Tyranny. The question that has always mattered is not whether power flows from the left or the right, but whether it is dispersed or concentrated, checked or unchecked, accountable or absolute.

Not every revolution ends in tyranny. Some movements achieve genuine constitutional transformation, dispersing power rather than concentrating it, and the conditions that produce this outcome rather than its opposite are among the central questions this series addresses. But every revolution, without exception, creates the conditions under which radicals can gain influence. The disruption of existing institutions, the collapse of legal norms, the polarization of factions, the cult of urgency that suspends ordinary deliberation, all of these open spaces for those who were previously marginal. And it is the radical, on both left and right, who is positioned to fill that space, because the radical alone has already committed to the use of political violence and therefore faces no threshold that others must cross before acting.

It is these revolutionary radicals, the utopians of both extremes who believe that a perfect society lies just beyond the present bloodshed, with whom this volume is principally concerned. Whether they march under the banner of proletarian liberation or national purification, they share a common intellectual ancestry in the Platonic conviction that the enlightened few are entitled, indeed obligated, to remake the many by force. And they share a common political trajectory: from the margins of a destabilized society, through the seizure of revolutionary momentum, to the concentration of power that extinguishes the very liberties their movements claimed to champion.

The circle closes. It always closes. The only question is whether constitutional institutions prove strong enough, and citizens virtuous enough, to hold the line at the base before the ascent begins.

The revolutionary terrorism value chain: a framework for analysis[74]

To analyze revolutionary patterns systematically, this book employs the

---

[74] Thomas Myers, "Terrorist to Tyrant," *Homeland Security Affairs* 7, art. 6 (February 2011).

Revolutionary Terrorism Value Chain (RTVC) model, which maps recurring stages through which utopian revolutionary movements progress. While each revolution possesses unique characteristics shaped by culture, ideology, and historical circumstances, the RTVC framework reveals structural patterns that recur across vastly different contexts.

The model identifies six core stages plus three alternative branches: democratic transition, civil war, and coup d'état:

Stage 1: ideological development: Revolutionary movements begin with ideas that propose utopian alternatives. Whether secular or religious (Marxist, nationalist, jihadist), revolutionary ideology provides moral framework for violence, identifies enemies requiring elimination, and promises transformation through political action. Effective revolutionary ideology combines grievance (explaining injustice), vision (promising a better future), and legitimation (justifying violence against oppressors).

Stage 2: the vanguard: A radical core operationalizes ideology through clandestine organization and targeted violence. Russia's People's Will terrorist, France's *Enragés*, and Iran's Fadayan-e Islam all represent vanguards willing to use terror systematically to advance revolution. Vanguard terrorism serves a strategic purpose: demonstrating regime vulnerability, provoking overreaction that radicalizes moderates, eliminating rivals within the revolutionary movement, and maintaining discipline through fear. The vanguard stage is crucial because it transforms abstract ideology into organizational capacity.

Stage 3: mass civil unrest and revolution: Broad popular mobilization combines overt protest with clandestine vanguard action. Mass demonstrations provide political cover and legitimacy; vanguard violence creates crisis forcing regime response. This stage features general strikes, street battles, occupation of public spaces, and escalating confrontation. Regimes face impossible choices: violent suppression radicalizes moderates and may trigger military defection; accommodation signals weakness and emboldens revolution. Revolutions can succeed or fail during this stage based on whether security forces remain loyal or fragment.

Stage 4: revolutionary victory and consolidation of power: Overthrown regimes are often replaced by a moderate but weak provisional government; competing factions struggle for control. Moderates who led mass mobilization discover that vanguards control revolutionary violence and therefore post-revolutionary authority. The faction most willing to use terror against rivals consolidates power. Radical factions consolidate power through purges, terror, and ideological control. Revolutionary governments discovered that governing requires eliminating rivals; first external enemies (aristocrats, capitalists, imperialists), then internal enemies (moderates, deviationists, counterrevolutionaries).

Stage 5: tyranny: There is at some point a struggle for leadership and a supreme ruler, having eliminated all rivals, emerges. Terror becomes an administrative tool for maintaining revolutionary purity. This stage features show trials, labor camps, mass executions, and personality cults. Ironically, revolutionary regimes often become more oppressive than those they replaced, justifying repression through revolutionary necessity.

Stage 6: export of terror and expansion: Revolutionary regimes sponsor foreign insurgencies and export ideology, viewing revolution as universal mission requiring global triumph. The French Revolution exported revolutionary war throughout Europe; Soviet Russia sponsored communist movements worldwide; The Iranian Revolution funded Shi'a militias across the Middle East. Export serves multiple purposes: spreading ideology, creating buffer zones against counter-revolution, demonstrating revolutionary vitality, and justifying continued emergency rule domestically.

Stage 7: democratic transition (alternative path): Some revolutions achieve rare "off-ramp" to constitutional democracy rather than tyranny. This requires pre-existing constitutional traditions, robust civil society mediating between factions, elite restraint accepting constitutional limits, and external security allowing experimentation without paranoia. Tunisia's Arab Spring transition, The United Provinces break from the Hapsburg Empire, England's glorious revolution, and America's founding represent successful democratic transitions.

Stage 8: civil war (optional branch): Revolutionary situations may devolve into prolonged civil war when multiple factions possess military capacity and external powers support different sides. Syria's multi-sided conflict, Libya's militia warfare, and Russia's post-revolutionary civil war (1918-1921) illustrate this trajectory. Civil wars extend revolutionary violence indefinitely, destroy state capacity, and often end with military strongman restoring order through force.

Stage 9: coup d'état (optional branch): Coups typically occur during the consolidation of power phase. Though coups can occur independently, radicals have seized power through coups within larger revolutions, as the Jacobins, Napoleon, and the Bolsheviks all demonstrated.

Chapter 1 develops the framework in detail, applying it to the Russian and Iranian revolutions to demonstrate its analytical power. Subsequent chapters apply the RTVC model to the Arab Spring (chapter 2), examine how it illuminates the English, American, and French revolutions (chapter 3), and explore contemporary American polarization through this lens (chapter 4).

The RTVC framework proves essential for understanding upheavals in our current era. It explains why revolutionary movements at home, whether left or right, follow predictable patterns toward tyranny.

The argument in brief

This *Revolutions and Constitutions* is a three volume book advancing three core arguments: (1) revolutions follow recurring patterns across ideologies and cultures, and the RTVC framework helps identify those patterns, their warning signs, and their threat to constitutional order; (2) the fundamental political divide is not left versus right but centralization versus decentralization, since revolutions tend to concentrate power while stable constitutional orders preserve liberty by dispersing it; and (3) legitimacy cannot be created through revolutionary violence or foreign intervention but must arise from indigenous constitutional processes that secure durable popular consent.

Revolutions do not emerge in a vacuum; they arise from the intersection of ideas, grievances, and institutional failures. Volume I of *Revolutions and Constitutions* lays the conceptual and historical groundwork for understanding how revolutions erupt, what animates them, and how their legacies shape, or fail to shape, legitimate constitutional orders. The centralization versus decentralization framework described in the introduction finds its first application here: revolutions are contests over sovereignty, who holds power, how it is legitimated, and whether it will be centralized in the hands of a revolutionary vanguard or dispersed through constitutional limits.

The RTVC model, introduced in chapter 1, maps how movements progress from ideological development through vanguard terrorism, mass mobilization, revolutionary victory, consolidation of power, and, too often, tyranny. The historical cases in chapters 2 and 3 demonstrate both the universality of these patterns and the critical variables that determine whether revolutions produce constitutional republics or revolutionary dictatorships. Chapter 4 brings this analysis into contemporary America, revealing that even stable constitutional democracies are not immune to revolutionary dynamics when centralization erodes mediating institutions and trust.

Chapter 1 — Theoretical Framing introduces the core conceptual architecture used throughout the book. Drawing on diverse traditions, ranging from classical revolutionary theory to modern frameworks like the RTVC, this chapter defines the stages and conditions that characterize revolutionary movements. It compares prevailing theories from scholars such as Goldstone, Tilly, Skocpol, and Chenoweth, offering readers an analytical lens to interpret historical and contemporary upheavals.[75]

The RTVC framework proves essential for understanding why some revolutions consolidate power tyrannically (Russia, Iran) while others achieve constitutional balance (the United States).[76] Jack Goldstone's complementary three-phase theory (state breakdown, struggle for power, state reconstruction) provides additional analytical depth.[77] Together, these frameworks enable systematic comparison across the cases that follow.

Chapter 2 — Justice and Jihad turns to the Arab Spring as a case study in modern revolutionary struggle, where calls for justice and reform intersected with Islamic political thought, authoritarian retrenchment, and civil war.[78] through analyses of Tunisia, Egypt, Libya, and Syria, this chapter applies the RTVC framework to map each country's unique trajectory, while also interrogating concepts of just war and Islamic legal reasoning.[79] the result is a sobering exploration of how revolutionary hope can be subverted by sectarianism, militarization, and competing claims to moral legitimacy.

The Arab Spring shows that revolutionary dynamics can follow familiar patterns even in Islamic political contexts. It also highlights variables the RTVC cannot fully capture, especially the strength of civil society and the capacity of preexisting institutions. These uprisings were further complicated by rival claims to legitimacy rooted in Arab nationalism and political Islam, each

---

[75] Jack A. Goldstone, *Revolution and Rebellion in the Early Modern World* (Berkeley: University of California Press, 1991); and idem, "Toward a Fourth Generation of Revolutionary Theory," *Annual Review of Political Science* 4 (2001): 139-187.; Charles Tilly, *From Mobilization to Revolution* (Reading, MA: Addison-Wesley, 1978).; Theda Skocpol, *States and Social Revolutions: A Comparative Analysis of France, Russia, and China* (Cambridge: Cambridge University Press, 1979).; Erica Chenoweth and Maria J. Stephan, *Why Civil Resistance Works: The Strategic Logic of Nonviolent Conflict* (New York: Columbia University Press, 2011).

[76] The Revolutionary Terrorism Value Chain (RTVC) framework is developed in detail in Chapter 1 of this book.

[77] Goldstone, "Toward a Fourth Generation," 145-151, outlines the three-phase theory: State Breakdown, Struggle for Power, and State Reconstruction.

[78] On the Arab Spring, see Marc Lynch, *The Arab Uprising: The Unfinished Revolutions of the New Middle East* (New York: PublicAffairs, 2012); and James L. Gelvin, *The Arab Uprisings: What Everyone Needs to Know*, 2nd ed. (Oxford: Oxford University Press, 2015).

[79] *Khurūj* (خروج) refers to armed rebellion against Muslim rulers. Classical Islamic jurisprudence generally prohibited such rebellion except under strictly limited conditions. See Khaled Abou El Fadl, *Rebellion and Violence in Islamic Law* (Cambridge: Cambridge University Press, 2001).

representing a different path a revolutionary outcome. Tunisia's relative success in avoiding both tyranny and civil war, and in achieving the rare RTVC off-ramp to democratic transition, owed much to a civil society strong enough to mediate among competing factions and block the consolidation of power. By contrast, Libya's collapse into state failure, Egypt's reversion to military rule through coup, and Syria's descent into civil war each illustrate the consequences of revolutions despite the hope for change.

The chapter's theological analysis of *jihad* and *khurūj* (rebellion against Muslim rulers) reveals how classical Islamic jurisprudence grappled with the same questions Protestant reformers confronted: when does obedience to unjust authority conflict with obedience to God? Sayyid Qutb's radical reinterpretation lowering the threshold for declaring rulers apostates and transforming rebellion from rare exception to permanent revolutionary duty, echoes the Jacobin and Bolshevik pattern of revolutionary vanguards claiming special enlightenment that justifies terror.[80] The Arab Spring's ultimate failure across most of the region demonstrates that revolutionary terrorism produces tyranny regardless of the ideology wielded, whether Islamist or secularist, nationalist or jihadist.

Chapter 3 — Revolutions and Constitutions shifts focus to the Atlantic world, examining the English, American, and French revolutions. These cases highlight the complex relationship between revolutionary violence and constitutional development. The chapter traces the ideological roots of constitutionalism in English common law, puritan theology, Enlightenment liberalism, and classical republicanism, exploring how each tradition informed both revolution and the subsequent struggle to institutionalize power.[81] By comparing the divergent outcomes of these revolutions, whether reform, republic, and terror, the chapter raises questions about the conditions under which revolutions give birth to lasting political orders.

All three revolutions receive systematic RTVC analysis, revealing crucial

---

[80] Sayyid Qutb, *Milestones* (New Delhi: Islamic Book Service, 2006 [1964]). On Qutb's revolutionary ideology, see John Calvert, *Sayyid Qutb and the Origins of Radical Islamism* (New York: Columbia University Press, 2010).

[81] On English common law and constitutionalism, see J.G.A. Pocock, *The Ancient Constitution and the Feudal Law* (Cambridge: Cambridge University Press, 1987); and Glenn Burgess, *The Politics of the Ancient Constitution* (University Park: Pennsylvania State University Press, 1993).; On Enlightenment liberalism and republicanism, see Bernard Bailyn, *The Ideological Origins of the American Revolution*, enlarged ed. (Cambridge, MA: Harvard University Press, 1992).; On Puritan political theology, see Michael Walzer, *The Revolution of the Saints: A Study in the Origins of Radical Politics* (Cambridge, MA: Harvard University Press, 1965).

patterns. The English civil war demonstrates that even successful revolutions can descend into tyranny (Cromwell's protectorate) before finding a constitutional off-ramp (the Glorious Revolution of 1688).[82] The American revolution shows that tyranny can be avoided entirely through moral restraint (Washington refusing kingship) and constitutional design (federalism, separation of powers, Bill of Rights).[83] The French revolution confirms the RTVC's tragic arc: ideological extremism (Rousseau's romantic utopianism), vanguard terror (Jacobins), revolutionary consolidation (committee of public safety), tyranny (reign of terror), and export (Napoleonic conquests).[84]

The comparative analysis reveals why outcomes diverged so dramatically. England and America preserved constitutional traditions and religious foundations that emphasized moral restraint and institutional continuity. France attempted total destruction of the *ancien régime*, rejecting all traditional authority in favor of revolutionary utopianism. The Protestant view that man is flawed and requires constitutional limits produced stable republics; the Rousseauian view that man is good but corrupted by bad institutions produced the terror.[85] Americans saw their revolution as triumphing through economic and political decentralization. From the French revolution came the political terminology of left and right, the conscious use of revolutionary terror as political tool, and the utopian motivation that traces from the Jacobins through the Bolsheviks to contemporary jihadists.[86]

Chapter 4 — Reciprocal Radicalism brings the analysis into the twenty-first century, exploring the escalating ideological polarization and street-level confrontations within the United States. From the Weather Underground to antifa, and from Aryan Nations to the Patriot Front, this chapter situates radical

---

82 On Cromwell's Protectorate (1653-1658), see Barry Coward, *Oliver Cromwell* (London: Longman, 1991).; On the Glorious Revolution, see Steve Pincus, *1688: The First Modern Revolution* (New Haven: Yale University Press, 2009).

83 On Washington's refusal of kingship and voluntary relinquishment of power, see Ron Chernow, *Washington: A Life* (New York: Penguin Press, 2010), especially 530-534 on the Newburgh Conspiracy and 784-794 on his retirement after two terms.

84 Jean-Jacques Rousseau, *The Social Contract*, trans. Maurice Cranston (London: Penguin Books, 1968). On Rousseau's influence on the French Revolution, see Carol Blum, *Rousseau and the Republic of Virtue: The Language of Politics in the French Revolution* (Ithaca: Cornell University Press, 1986).

85 John Locke, *Two Treatises of Government*, ed. Peter Laslett (Cambridge: Cambridge University Press, 1988).

86 On the Reign of Terror, see R.R. Palmer, *Twelve Who Ruled: The Year of the Terror in the French Revolution* (Princeton: Princeton University Press, 1941); and Simon Schama, *Citizens: A Chronicle of the French Revolution* (New York: Knopf, 1989).

movements within a broader pattern of revolutionary behavior. It explores how political violence, disinformation, and symbolic conflict reflect deeper fractures in the American constitutional order and what these dynamics reveal about the health of modern liberal democracies.

The Weimar Germany parallel proves disturbingly apt: radical left and radical right factions engaged in street violence, each justifying their actions as defensive responses to the other's provocations, while moderate democratic institutions eroded.[87] the Weather Underground's transformation from students for a democratic society (SDS) into a terrorist organization illustrates how radical factions emerge from broader social movements when institutional reform seems impossible.[88]

Contemporary America has not reached Weimar's precipice, but warning signs are evident. Trust in institutions has collapsed. Both left and right increasingly view elections as existential contests rather than routine democratic processes. Reciprocal radicalization, each side's extremism justifying the other's, accelerates polarization. Social media amplifies outrage and creates echo chambers impervious to contrary evidence. The mediating institutions that once moderated conflict, churches, civic organizations, local governments, responsible media, have weakened precisely when their moderating influence is most needed.

Together, these chapters offer an integrated foundation for understanding the anatomy of revolutionary terrorism and how ideas ignite insurgency. The RTVC framework reveals a tragic pattern: revolutionary movements that resort to terrorism rarely give up the use of terror afterward. The radical vanguard that seizes the state becomes the new tyranny; often more oppressive than the regime it replaced. Yet the framework also identifies rare off-ramps: the United Provinces, England's glorious revolution, America's constitutional founding, Tunisia's negotiated transition. What distinguishes success from failure?

This volume's comparative analysis suggests several critical variables:

Pre-existing constitutional traditions matter immensely. England's centuries of common law and parliamentary precedent provided a framework for limiting royal power without destroying all traditional authority. America

---

[87] On Weimar Germany and political polarization, see Eric D. Weitz, *Weimar Germany: Promise and Tragedy* (Princeton: Princeton University Press, 2007); and Peter Fritzsche, *Germans into Nazis* (Cambridge, MA: Harvard University Press, 1998).

[88] On the Weather Underground, see Bryan Burrough, *Days of Rage: America's Radical Underground, the FBI, and the Forgotten Age of Revolutionary Violence* (New York: Penguin Press, 2015).

inherited this English constitutional tradition and improved upon it through federalism. France, having destroyed comparable constitutional culture, attempted to build from scratch based on abstract reason, with catastrophic results.

Religious and moral foundations that emphasize restraint, humility, and institutional continuity prove essential. The puritan conviction that all men are sinners, including revolutionary leaders, produced Washington's voluntary relinquishment of power. The Enlightenment conviction that well intentioned revolutionaries could perfect society through reason produced Robespierre's terror.

Geographic security allows constitutional experimentation without paranoia. America's ocean barriers provided breathing room; France's vulnerability to European monarchies produced siege mentality and revolutionary extremism.

Dispersed social power in civil society creates alternatives to state control. Tunisia's robust labor unions and professional associations mediated between Islamists and secularists, preventing either from monopolizing power.[89] Libya's tribal fragmentation, lacking comparable civic institutions, produced warlordism after Gaddafi's fall.

Elite restraint determines whether revolutionary victory produces constitutional transition or vanguard dictatorship. When revolutionary leaders voluntarily accept institutional limits, as American founders did, constitutional stability becomes possible. When they view constitutional constraints as obstacles to revolutionary transformation, as Jacobins, Bolsheviks, and Khomeinists did, tyranny follows.

These variables operate within the larger centralization-decentralization dynamic. Revolutions that disperse power through constitutional federalism, separation of powers, and civil society pluralism achieve stability. Revolutions that concentrate power in revolutionary committees, supreme leaders, or ideological vanguards produce tyranny regardless of stated intentions.

The journey through Volume I establishes that revolutionary patterns recur across time, culture, and ideology. The same dynamics that produced the French terror recurred in Russia, Iran, and Libya. The same constitutional

---

[89] On Tunisia's civil society and democratic transition, see Monica Marks and Omar Belhaj Salah, "Uniting for Tunisia? Ennahda and Nidaa Tounes in the Country's 2014 Elections," *Brookings Doha Center Analysis Paper* 12 (January 2015); and Youssef Cherif, "Tunisia's Risky Hardline Approach to Terrorism," *Carnegie Endowment for International Peace* (April 2, 2015).

wisdom that stabilized England and America offers lessons for contemporary transitions. The question is whether policymakers, revolutionaries, and citizens will heed these lessons or repeat history's tragic cycles.

CHAPTER 1

THEORETICAL FRAMING

A movement is pioneered by men of words, materialized by fanatics and consolidated by men of action.
— Eric Hoffer, *The True Believer*

This chapter is based on my 2011 essay, published at the height of the Arab Spring. Titled *Terrorist to Tyrant*, the article asked a question: What happens when terrorist groups achieve their political goals? A comparative analysis of the Russian and Iranian Revolutions and Hezbollah's rise in Lebanon led to the development of the Revolutionary Terrorism Value Chain (RTVC), a model tracing the evolution of violent revolutionary movements into governing regimes.

This volume builds on and expands that framework. It situates the RTVC within a broader literature on revolutions and state power, engaging especially with Jack Goldstone's comprehensive theory of revolution. To clarify the scope, this book uses the terms terrorism, insurgency, and revolution as interconnected phases of conflict, each operating at different levels of war:

- Terrorism is defined here as "the calculated use of violence outside internationally accepted bounds of civil law and conventional military conduct in pursuit of political or social objectives."[90] It is a tactical activity designed to provoke, disrupt, or attract attention.

- Insurgency involves sustained, organized resistance to a government or occupying force, using both violent and political means. It functions at the operational level, linking tactical actions to strategic goals.

- Revolution refers to a strategic-level transformation: "The forcible overthrow of a government through mass mobilization (whether

[90] Mark Burgess, "Terrorism: The Problems of Definition" (*Center for Defense Information*, August 1, 2003).

military or civilian or both) in the name of social justice, to create new political institutions."[91]

A distinction is also made between revolutions, which seek to overthrow a domestic government, and revolts, which resist external occupation. Furthermore, the doctrinal term irregular warfare encompasses terrorism, insurgency, and revolution, in contrast to conventional war, which occurs between sovereign states employing regular armed forces.[92]

Understanding how terrorist or insurgent movements evolve into ruling regimes requires a framework that captures both the political and psychological mechanisms of revolutionary change. Before introducing the RTVC in detail, we turn to one of the most influential models in the field: Jack Goldstone's theory of revolution.

Goldstone's Theory of Revolution

Goldstone outlines three core phases of revolution:

1. State Breakdown: Triggered by a crisis—often fiscal or military—in combination with elite defection and the ability to mobilize the masses.
2. Struggle for Power: Competing revolutionary coalitions seek control; moderate governments often replace the old regime but falter, allowing radicals to seize power.
3. State Reconstruction: The nature of the revolutionary government is dependent on the background of revolutionary leaders, the state of the pre-existing political and economic system, and international context.

The first phase, State Breakdown, occurs when a regime faces three converging conditions: a deep crisis, elite defection, and mass mobilization potential. Political dysfunction, what Jonathan Hackett terms "sovereign dysfunction," erodes state legitimacy when citizens' basic expectations are unmet, often due to fiscal collapse or political deadlock.[93] When a segment of the elite turns against the regime, and large-scale protests emerge, the state's capacity to maintain order declines, setting the stage for open conflict.

The second phase, Struggle for Power, unfolds when revolutionaries organize coalitions strong enough to challenge the regime. Success depends on

---

[91] Jack A. Goldstone, *Revolutions: A Very Short Introduction* (Oxford: Oxford University Press, 2014), 4.

[92] Thomas Myers, "Irregular War is Revolutionary," (CTX Journal, Spring 2021), 7-8.

[93] Hackett, Jonathan, *Theory of Irregular War,* Kindle Edition (McFarland & Company Jefferson, NC 2024), 14.

unity among insurgent groups and the state's failure to divide or suppress them. Often, a "honeymoon" period follows regime collapse, during which a provisional government must act quickly to govern effectively and address core grievances. History shows that failure in this phase frequently leads to a radical takeover, as with the Jacobins in France or the Bolsheviks in Russia.

The final phase, State Reconstruction, involves the consolidation of new institutions and authority. The outcomes of this phase depend heavily on the background of revolutionary leaders, the state of the pre-existing political and economic system, and the international environment.

While Goldstone's model offers a comprehensive framework for revolution in general, the Revolutionary Terrorism Value Chain (RTVC) focuses specifically on radical or terrorist actors within that process, those who begin on the margins, use violence to advance their cause, and often emerge as dominant forces in the post-revolutionary state.

Two rival frameworks deserve acknowledgment. Theda Skocpol's structural theory emphasizes international pressures and agrarian class structures over ideological agency, arguing that revolutions are not "made" by revolutionaries but result from structural crises beyond any actor's control.[94] The RTVC diverges from Skocpol in emphasizing vanguard agency and ideological motivation, treating structural conditions as necessary but not sufficient. Charles Tilly's resource mobilization model focuses on how groups organize collective action rather than on revolutionary stages, emphasizing the relationship between state repression and popular mobilization.[95] The RTVC incorporates elements of Tilly's analysis in its attention to vanguard formation (Stage 2) and mass mobilization (Stage 3) but adds the trajectory toward tyranny that Tilly's more process-oriented model does not predict. Neither Skocpol nor Tilly adequately accounts for the ideological dimension that drives revolutionary movements beyond pragmatic grievance toward utopian transformation, which is the RTVC's central concern.

A third framework, distinct from both Skocpol and Tilly, approaches revolutionary conflict from the military rather than the sociological tradition. Mao Tse-tung's theory of protracted war, developed through the Chinese Communist Revolution and systematized in his 1938 treatise *On Protracted War*, offers perhaps the most influential operational framework for understanding how a revolutionary movement translates ideological commitment and mass

---

94 Skocpol, Theda. *States and Social Revolutions: A Comparative Analysis of France, Russia, and China.* Cambridge: Cambridge University Press, 1979.

95 Tilly, Charles. *From Mobilization to Revolution.* Reading, MA: Addison-Wesley, 1978.

mobilization into the seizure of state power.[96] Where Skocpol and Tilly analyze the structural and organizational conditions that make revolution possible, Mao addresses the strategic problem of how a militarily inferior movement defeats a superior state over time.

Mao's central insight was that political organization, military action, and information operations are inseparable instruments of a single strategic effort. He identified three sequential but overlapping phases through which a revolutionary movement progresses from marginal insurgency to state power.

The first phase, strategic defensive, corresponds to the period in which the state holds the initiative and the revolutionary movement is building its foundation. Organizers work both overtly and clandestinely, constructing cells and networks, infiltrating government institutions and civil organizations, and conducting information operations designed simultaneously to build popular support for the revolutionary cause and erode confidence in the existing regime. The movement at this stage avoids direct military confrontation it cannot win, instead manufacturing the appearance of spontaneous popular discontent through organized strikes, protests, and boycotts.

The second phase, guerrilla warfare, begins when the movement has sufficient organization and popular support to apply direct pressure on the state. Sabotage, terrorism, and targeted propaganda intensify. The movement seeks international patrons, establishes base areas beyond effective state control, and begins tactical military operations. As momentum builds, the insurgency consolidates territorial control, proclaims a counter-government, and systematically expands both its military capacity and its political administration.

The third phase, war of movement, becomes possible only when the revolutionary force has grown sufficient conventional capacity to contest the state's military power directly. Guerrilla and terror operations continue, but a conventional military dimension is added, at which point the conflict has become a civil war in the full sense. Victory in this phase delivers political

[96] Mao Tse-tung. "On Protracted War." In *Selected Works of Mao Tse-tung*, vol. 2, 113–194. Peking: Foreign Languages Press, 1967. Marxists Internet Archive. https://www.marxists.org/reference/archive/mao/selected-works/volume-2/mswv2_09.htm.; Mao Tse-tung. *On Guerrilla Warfare.* Translated and with an introduction by Samuel B. Griffith II. New York: Frederick A. Praeger, 1961.

control of the state and, in the Maoist model, the wholesale replacement of the existing socioeconomic order.

Several qualifications are necessary. Actions at the tactical, operational, and strategic levels occur simultaneously within each phase; phases themselves overlap and interpenetrate rather than succeeding one another cleanly. More importantly, Mao's model describes one trajectory among several possible outcomes. A regime may collapse before the third phase develops (as in Tunisia), a coup may interrupt the insurgent progression (Egypt), or the movement may be forced by battlefield reversal to revert to an earlier phase (as the Islamic State was compelled to do following its territorial losses after 2017). The framework is best understood not as a deterministic sequence but as a map of branches and sequels: the likely path absent disruption, with alternative trajectories always available depending on the relative capacity and choices of the competing actors. Mao's theory will receive fuller treatment in Volume II, where his analysis of external support as a decisive variable in revolutionary conflict receives the attention, it warrants.

Mao's three-phase framework illuminates the operational mechanics of revolutionary escalation, but it does not fully account for the ideological engine that carried a victorious movement past liberation to tyranny. The three phases describe how a revolutionary force grows; they do not explain why radical vanguards, once in power, reliably betray the constitutional promises that justified their struggle. To address that question, this volume draws on a complementary analytical instrument developed from the discipline of intelligence analysis rather than military theory.

The Revolutionary Terrorism Value Chain (RTVC)

The Revolutionary Terrorism Value Chain is derived from David Rubin's "Transforming Intelligence Using Industry Best Practices-The Terrorism Value Chain."[97] Value Added methodology was developed in the public commercial sector to increase the effectiveness of manufacturing and purchasing processes. Also known as Value Chain Analysis, this methodology breaks down a process

[97] David Rubin, Maya Vidich, David Kletter, and Sharon Russ. "Transforming Intelligence Using Industry Best Practices - The Terrorism Value Chain," NOTE: This paper is no longer available online.

to its component parts. Each part is examined to determine if that part can be eliminated or enhanced to augment the overall efficiency of the entire process.

Because this method breaks terror operations into steps, the analyst can determine at what stage a terrorist organization is in their operational cycle. From start to finish the Terrorism Value Chain breaks down terror operations into the steps necessary to produce an attack, from planning to gathering resources to carrying out the operation. The Terrorist Value Chain consists of the following: reconnaissance, target selection, operational planning, team selection, specialized training, resourcing, rehearsal, final preparations and movement, and the attack. These steps can overlap, may not be sequential, and yet provide multiple points at which the analyst might get inside the terrorist decision cycle. Value Chain Analysis is a tool to identify where and how to task collection and to identify indicators of impending terror operations.

The RTVC melds Rubin's Terrorism Value Chain, applied at a macro level, with revolution theory. Many revolutionary movements begin with ideological rejection of the status quo and progress through increasingly complex stages of violence and political control. While the terms "terrorism" and "terrorist" dominate discourse, they describe a tactic not a motivation. Groups use terror as a means to an end: regime change.

The RTVC functions as a diagnostic tool rather than a causal model. It identifies a recurring pattern that revolutionary movements have historically exhibited, but it does not claim that all movements must follow this sequence or that progression through one stage inevitably produces the next. The RTVC was specifically developed to analyze revolutionary movements which resort to terrorism to achieve their goals. Its analytical value lies in pattern recognition: when analysts observe a movement entering a particular stage, historical precedent suggests which dynamics are likely to emerge.

The framework is most powerful when applied to movements that espouse utopian ideologies and pursue state power through organized violence. Cases where the model's fit is imperfect, such as the American Revolution's avoidance of tyranny or Tunisia's democratic transition, are as analytically important as cases where the pattern holds, because they reveal the conditions under which the cycle can be broken.

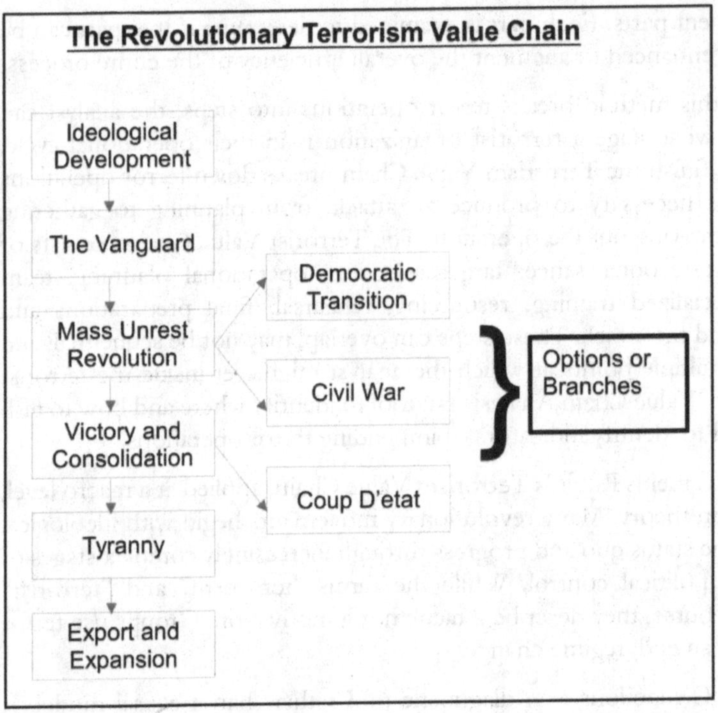

1. Ideological Development

Revolution begins in the realm of ideas. Movements articulate grievances, envision alternative futures, and justify extraordinary means.

A compelling revolutionary ideology typically includes:

- A diagnosis of the current regime's failure or injustice
- A vision of a transformed society (political, religious, economic)
- A justification for extraordinary means, including terroristic violence

2. The Vanguard

A small, radical core operationalizes the ideology forming a clandestine organization. The vanguard, akin to Lenin's party model or Che Guevara's *foco,* conducts targeted acts of terror not only to weaken the regime, but also to provoke a disproportionate response to generate sympathy for the movement and expand the revolutionary base.

Examples include:

- The Weather Underground (U.S.)
- Red Brigades (Italy)

- Narodnaya Volya (Russia)
- Al Qaeda (Middle East)

State responses may range from repression to reform. If repression is excessive and poorly managed, it can inadvertently amplify the insurgents' legitimacy.

3. Mass Civil Unrest and Revolution

With increased popular engagement, revolutionary ideology transitions from fringe doctrine to mass movement. The radical group continues terrorist activities; however, the group maintains a separation between the action cells responsible for sabotage, assassination, and kidnappings and the broader political party formed out of the disaffected population. The overt political party employs legitimate means of civil protest, including protest marches, general strikes, and sit-ins. Catalyzing events, economic crises, political scandals, or violent repression, mobilize broader sectors of society. The movement now combines:

- Overt political parties advocating reform or regime change
- Coalition-building to create legitimacy and mass appeal
- Infiltration and co-opting coalition partners
- Clandestine cells conducting sabotage, assassinations, infiltrating the regime, and provoking riots
- Mass protest actions (e.g., protest marches, general strikes, sit-ins)

Violent and nonviolent tactics often coexist. The line between spontaneous unrest and centrally orchestrated escalation is deliberately blurred by revolutionary leaders. As support grows, so does the radial groups' ability to field broader coalitions and challenge state authority. If the unrest continues to grow an armed insurgency may develop.

Again, the State may choose to respond with repression or reformation. If the government can address the underlying grievances, the perceived need for revolution can be removed. The must be a response to lawless activities. The failure to provide law and order will become yet another grievance to be exploited by the radical. Security forces have a fine line to walk. Overreaction too, can exacerbate the situation.

4. Revolutionary Victory and Consolidation of Power

When revolutionary forces succeed in toppling the regime, a power vacuum opens, and competing factions often clash over the post-revolutionary order. Initially, a moderate provisional government may take control, but history shows that the most organized and ruthless faction often seizes power.

Key mechanisms of consolidation include:

- Purges of rivals, including former allies
- Centralization of authority under a single party or leader
- Control of security forces
- Use of state terror to suppress opposition
- Mass arrests and censorship
- Creation of new institutions that reinforce ideological control

Examples include:

- The Bolsheviks eliminating Mensheviks and Social Revolutionaries (Russia)
- The Jacobins executing Girondins (France)
- Khomeini's purging of liberals and Marxists after 1979 (Iran)

5. Tyranny

As some point there is a power struggle within the radical group and a sole, supreme leader will assert total control. Power becomes increasingly centralized, opposition is criminalized, and violence becomes institutionalized.

Common features include:

- State terror: systematic violence against dissenters
- Cult of personality: near-deification of the supreme leader
- Elimination of legal opposition: political pluralism is extinguished
- Ideological indoctrination: through education, media, and surveillance

Historical precedents:

- Robespierre's execution of Danton
- Stalin's assassination of Trotsky
- Stalin's Great Terror
- Mao's Cultural Revolution
- The Iranian 1988 prison massacres

At this stage, the revolution often abandons its original promises of justice or equality in favor of stability through repression.

6. Export of Terror and Expansion

Having consolidated power, some regimes seek to export revolution. This may serve strategic, ideological, or defensive purposes.

Methods of export include:

- Military conquest;
- Funding and arming foreign insurgent groups;
- Providing sanctuary or training camps;
- Using state intelligence agencies to support proxy terrorism;
- Promoting revolutionary ideology through media or diplomacy.

Examples:

- Napoleon's conquests
- Soviet Expansion into Eastern Europe
- The Soviet Comintern's support for Marxist terrorists in Western Europe and revolutions abroad;
- Iran's sponsorship of Hezbollah and other Shi'a militant networks;
- Cuba's support for insurgencies in Latin America and Africa.

Exporting terror extends the revolution's influence, but also invites international isolation, sanctions, or intervention.

7. Alternative Path: Off-Ramp to Democratic Transition

Some revolutions or regimes de-escalate rather than radicalize. This off-ramp may involve:

- Negotiated transition to democratic governance
- Power-sharing agreements between former regime and opposition
- Voluntary exit of authoritarian leaders under domestic or international pressure

Success depends on:

- Military/Security Force neutrality or support for reform
- Institutional readiness for elections
- Presence of moderate political leadership

Examples:

- Tunisia (2011–2014): a negotiated democratic transition
- Poland (1989): peaceful regime transformation
- Spain post-Franco: constitutional monarchy and democratization

However, many revolutions fail to take this route, either due to the dominance of radicals or structural weaknesses in civil society and state institutions.

8. Civil War (Optional Branch)

Not all revolutions escalate into civil wars, but when insurgent forces achieve military parity and control territory, prolonged armed conflict may follow. This is often referred to as the "war of movement" phase in Maoist doctrine and has occurred in major revolutions such as those in Russia (1917–1921), China (1927–1949), and Syria (2011-2025).[98]

David Armitage, in *Civil War: A History in Ideas*, describes internal conflicts within a single state in which both sides claim legitimacy, possess organized military forces, and control distinct geographic areas.[99] Both civil war and revolution fit that criterion; civil war is distinguished when the rebels achieve rough military parity with the regime and gains control of territory in effect splitting the country in two or more entities.[100]

The likelihood of civil war depends on:

- The strength and cohesion of the regime, especially the military
- The degree of revolutionary military organization
- The presence of foreign support for either side
- The failure of elite defections or negotiated reforms

Civil war is more likely when state authority fractures and foreign support polarize the conflict.

9. Coup d'état (Optional Branch)

When Coups occur, it is often during the consolidation of power phase. In the *State against the State: Theory and practice of the Coup d'état*, Eric Carlton defines a coup as a usually violent struggle for political power, which may not directly affect the population.[101] Coups are distinguished from revolutions, in that the perpetrators are members of the regime that they wish to overthrow.[102]

A coup may be carried out independent of any larger revolutionary struggle; however, it is worth noting the following examples of radicals and tyrants seizing power through coups.

---

[98] Although the Assad regime fell and a new regime is in power violence continues to be widespread and, as of this writing, it is not clear the fighting is over.

[99] David Armitage, *Civil War: A History in Ideas*, yale university press, 2017, 31.

[100] Ibid, 131.

[101] Eric Carlton, *The State against the State: Theory and practice of the Coup d'état*, Scholar press, Aldershot England, 1997, 6.

[102] Ibid, 1.

- France: By both the Jacobin and by Napoleon
- Russia: The Bolshevik's October Revolution's
- Egypt: Al Sisi's ouster of Morsi and the Muslim Brotherhood.

CASE STUDIES: THE RUSSIAN AND IRANIAN REVOLUTIONS

The following cases studies, adapted from *Terrorist to Tyrant,* serve as a primer of the RTVC's analytical application.

The Russian Revolution: Communism and State Terror

Ideological Development

The ideological origins of the Russian Revolution lay not only in Karl Marx's ideas of class struggle, but also in critiques of the Czarist autocracy by figures such as Mikhail Bakunin and Sergey Nechayev. Bakunin envisioned an anarchist society achieved through mass uprising, while Nechayev's *Catechism of a Revolutionary* promoted total commitment to revolutionary goals, including the elimination of all moral constraints.[103] Their ideas influenced underground revolutionary networks, particularly those that saw terrorism not only as a political tactic but as a moral imperative against tyranny.

The Vanguard – Small Group Terror – Clandestine Organization

In the late 19th century, revolutionary cells such as Narodnaya Volya (People's Will) emerged, adopting a strategy of targeted violence against the ruling elite. They carried out a series of high-profile assassinations, including the successful killing of Czar Alexander II in 1881.[104] These groups operated in secrecy, relied on ideological zeal, and believed that a vanguard of committed revolutionaries could catalyze a broader uprising.

Mass Civil Unrest and Revolution

The revolutionary groundwork laid by the clandestine vanguard erupted into mass mobilization in the early 20th century. The 1905 Revolution, triggered by events such as the Bloody Sunday massacre, brought strikes, peasant uprisings, and the formation of workers' councils known as soviets. The

[103] Mikhail Bakunin, *Statism and Anarchy* (Cambridge University Press, 1990); Sergey Nechayev, "Catechism of a Revolutionary," in *The Revolutionary Catechism and Selected Writings* (International Publishers, 1971).

[104] Franco Venturi, *Roots of Revolution: A History of the Populist and Socialist Movements in Nineteenth Century Russia* (University of Chicago Press, 1983), 601–615.

uprising was suppressed but the unrest festered and spread more widely. The final crisis was precipitated by financial difficulties and battlefield setbacks during the First World War. The February Revolution of 1917 led to the abdication of Czar Nicholas II, as discontent among soldiers, industrial workers, and peasants coalesced into a nationwide rebellion.[105] These uprisings were marked by spontaneous mass participation rather than centralized revolutionary leadership. The radicals were, however, organized and able to gain control of the wider movement.

Option: Civil War

Following the Bolshevik seizure of power in October 1917, Russia plunged into a brutal civil war between the Red Army, led by Lenin and later Trotsky, and the anti-communist White forces, a coalition of monarchists, liberals, and foreign-backed interventionists. The war devastated the countryside, militarized politics, and entrenched the Bolsheviks' reliance on coercive methods of control.[106]

Revolutionary Victory and the Coup d'État

While a Provisional Government, formed after the February Revolution and led by moderate reformers like Alexander Kerensky, attempted democratic governance, it proved incapable of ending involvement in World War I or restoring domestic order.[107] In what was effectively a coup d'état, the Bolsheviks stormed the Winter Palace in October 1917, dissolved the Constituent Assembly, and claimed exclusive authority to rule in the name of the Soviets.[108] Lenin justified this power grab as a necessary step toward proletarian dictatorship and the suppression of bourgeois sabotage.

Consolidation of Power

Once in power, the Bolsheviks launched a campaign to eliminate rivals across the political spectrum. The Red Terror, led by the Cheka (secret police), targeted former czarists, mensheviks, anarchists, and even socialist allies. After Lenin's death in 1924, Joseph Stalin emerged as the dominant figure through calculated manipulation of party machinery, exile or execution of rivals like Trotsky, and ideological purges that eliminated any dissent within the

---

105 Sheila Fitzpatrick, *The Russian Revolution* (Oxford University Press, 2001), 1–30.
106 Evan Mawdsley, *The Russian Civil War* (Pegasus Books, 2007), 55–90.
107 Alexander Rabinowitch, *The Bolsheviks Come to Power: The Revolution of 1917 in Petrograd* (Haymarket Books, 2004), 300–320.
108 Richard Pipes, *The Russian Revolution* (Vintage Books, 1991), 501–540.

Communist Party.[109] Stalin's consolidation of power formalized a cult of personality and transitioned the revolution into autocracy.

Tyranny

Under, first Lenin then Stalin, the revolutionary promise degenerated into tyranny. The regime orchestrated mass executions during the Great Purge, established a vast network of Gulag labor camps, and engineered famines such as the Holodomor in Ukraine to suppress resistance. World War II enabled Stalin to solidify control as the people rallied to the regime to protect the motherland. The Communist state became a totalitarian apparatus marked by surveillance, repression, and militarization.[110]

Export and Expansion

Following the consolidation of internal control, the Soviet Union turned outward, exporting revolution through the conquest of Eastern Europe and later through Cold War proxy wars. The Communist International, controlled by the Soviets, provided ideological and material support to insurgent and terrorist groups like the Red Brigades in Italy, Baader-Meinhof in Germany, and numerous Marxist movements across Africa, Asia, and Latin America. The USSR framed this expansion as part of a global struggle against imperialism.[111]

Off-Ramp: Democratic Transition (Foreclosed)

While the Provisional Government had offered a democratic alternative in 1917, its downfall foreclosed a peaceful transition. Once the Bolsheviks assumed power, any potential for pluralism was extinguished. Elections were dismissed as bourgeois formalities, and the one-party state became the institutional expression of revolutionary victory.[112] No democratic off-ramp emerged until the Soviet Union's collapse in 1991, by then, a different era and a different revolution.

---

[109] Robert Conquest, *The Great Terror: A Reassessment* (Oxford University Press, 2007), 79–120.

[110] Anne Applebaum, *Gulag: A History* (Anchor Books, 2004), 5–30; Timothy Snyder, *Bloodlands: Europe Between Hitler and Stalin* (Basic Books, 2010), 53–130.

[111] Christopher Andrew and Vasili Mitrokhin, *The World Was Going Our Way: The KGB and the Battle for the Third World* (Basic Books, 2005), 70–140.

[112] Orlando Figes, *A People's Tragedy: The Russian Revolution 1891–1924* (Penguin Books, 1996), 724–755.

RTVC Stage	Russian Revolution Summary
Ideological Development	Influenced by anarchist (Bakunin) and Marxist (Lenin) critiques of Tsarist autocracy.
The Vanguard	Narodnaya Volya and Bolsheviks; engaged in targeted assassinations and revolutionary planning.
Mass Civil Unrest & Revolution	1905 and 1917 revolutions driven by economic crisis, war exhaustion, and public protest.
Civil War (Optional Branch)	Red Army vs. White Army (1917–1921); Bolsheviks consolidate control through armed victory.
Revolutionary Victory	Provisional Governement takes power, led by the Menscheviks.
Coup d'État (Optional Branch)	October Revolution; Bolsheviks seize Winter Palace, dissolved the Provisional Government, and declared Soviet rule.
Tyranny	Upon Lenin's death Stalin consolidates power; Great Purge, forced collectivization, widespread repression.
Export of Terror	Soviet support for Comintern, Marxist insurgencies globally.
Democratic Transition	No off-ramp until USSR's collapse in 1991.

The Iranian Islamic Revolution: Martyrdom and Theocracy

Ideological Development

In the decades preceding the 1979 revolution, Ayatollah Ruhollah Khomeini articulated a radical reinterpretation of Shi'a theology that emphasized political sovereignty grounded in divine authority. His theory of *velayat-e faqih* (Guardianship of the Jurist) provided the religious justification for clerical rule, positioning the Islamic scholar as the rightful head of state. From exile, Khomeini denounced the Shah's regime as tyrannical, corrupt, and illegitimate, accusing it of Western subservience, moral decadence, and betrayal of Islamic principles.[113] His messages, smuggled via audio cassettes and letters, were widely disseminated by a clandestine clerical network and helped forge an ideologically unified opposition centered on Islamic governance and anti-imperialism.

The Vanguard

Before mass mobilization, groups like Fadayan-e Islam laid the groundwork for revolution by engaging in targeted political violence. Founded by Navab Safavi in the 1940s, the group assassinated secular politicians and intellectuals seen as enemies of Islam.[114] Later organizations such as Mujahedin-e Khalq (MEK) also used armed struggle and bombings to challenge the Shah's regime, though they diverged ideologically from Khomeini's vision. These vanguard movements contributed to a climate of resistance, demonstrating that the monarchy could be violently challenged.

Mass Civil Unrest and Revolution

By late 1978, public frustration with political repression, economic inequality, and the Shah's modernization efforts boiled over into nationwide protests and general strikes. The cycle of mourning and protest, based on the Shi'a 40-day commemoration tradition, created a feedback loop of escalating mobilization. The Black Friday massacre in Tehran (September 8, 1978), in which hundreds of protestors were killed by security forces, delegitimized the regime further and solidified the perception of the Shah as a tyrant.[115] Millions eventually took to the streets, effectively paralyzing the government.

---

[113] Ruhollah Khomeini, *Islamic Government: Governance of the Jurist* (Alhoda UK, 2002); Ervand Abrahamian, *A History of Modern Iran* (Cambridge University Press, 2008), 162–178.

[114] Nikki R. Keddie, *Modern Iran: Roots and Results of Revolution* (Yale University Press, 2006), 110–112.

[115] Said Amir Arjomand, *The Turban for the Crown: The Islamic Revolution in Iran* (Oxford

Option: Civil War

Unlike other revolutionary contexts, Iran narrowly avoided full-scale civil war. Instead, the Shah's power eroded rapidly as key elements of the armed forces began to defect or stand down. The upper echelons of the military, sensing the Shah's impending departure and the futility of violent repression, ultimately chose not to engage in a scorched-earth defense of the monarchy.[116] This allowed for a relatively swift collapse of the old order without prolonged internal conflict.

Off-ramp: Democratic Transition

In a last-ditch effort to prevent revolution, the Shah appointed Shahpour Bakhtiar, a moderate opposition leader, as Prime Minister in January 1979. Bakhtiar promised democratic reforms, political liberalization, and a constitutional monarchy. However, his government lacked legitimacy, was rejected by both Khomeini's supporters and the Shah's allies, and was quickly dismantled after the Shah fled the country.[117] Bakhtiar's fate underscored the weakness of liberal reform efforts in revolutionary climates dominated by absolutist alternatives.

Revolutionary Victory and Consolidation of Power

In February 1979, Khomeini returned to Iran to a hero's welcome and immediately sidelined the provisional government. The revolutionary coalition, initially diverse, was swiftly purged of liberals, Marxists, and moderate Islamists. The establishment of the Islamic Republic was ratified by referendum, followed by a new constitution enshrining clerical authority.[118] Khomeini's consolidation of power involved not only institutional restructuring but also an ideological transformation of Iranian society.

Option: Coup d'État

The post-revolution period featured a de facto clerical coup, as Khomeini used his position as Supreme Leader to eliminate rivals. Revolutionary courts conducted mass executions of political opponents, and paramilitary organizations like the Revolutionary Guard (IRGC) enforced the will of the

---

University Press, 1988), 94–110.

[116] Shaul Bakhash, *The Reign of the Ayatollahs: Iran and the Islamic Revolution* (Basic Books, 1984), 45–50.

[117] Abbas Milani, *The Persian Sphinx: Amir Abbas Hoveyda and the Riddle of the Iranian Revolution* (Mage Publishers, 2000), 368–372.

[118] Hamid Dabashi, *Theology of Discontent: The Ideological Foundation of the Islamic Revolution in Iran* (Transaction Publishers, 2006), 244–260.

new regime.[119] Internal dissent within the religious establishment was also suppressed, centralizing power in Khomeini's hands and subordinating all branches of government to the clerical elite.

Tyranny

The Islamic Republic transitioned quickly into a repressive theocracy. The regime-imposed Sharia law, restricted women's rights, curtailed freedom of speech, and censored the press. The newly formed Basij militia and the IRGC became key instruments of internal surveillance and political repression. Mass executions, particularly during the 1988 prison massacres of leftist dissidents, and the systematic targeting of minorities and secularists cemented the authoritarian character of the state.[120] Dissent was framed as apostasy or treason, legitimizing brutal crackdowns.

Export and Expansion

Iran's revolutionary mission did not stop at its borders. The Islamic Republic aggressively pursued the export of its ideological model, particularly to Shi'a populations in the Middle East. Its most significant creation was Hezbollah in Lebanon, which emerged in the early 1980s with Iranian training, funding, and religious guidance.[121] Iran also extended support to militant groups in Iraq, Syria, Bahrain, and Yemen, embedding its revolutionary principles in transnational networks. This "resistance axis" became central to Iran's foreign policy and regional strategy.

RTVC Stage	Iranian Revolution Summary
Ideological Development	Khomeini reinterprets Shi'a theology into political doctrine (velayat-e faqih); denounces Shah.
The Vanguard	Fadayan-e Islam conduct targeted assassinations and bombings.
Mass Civil Unrest &	1978–79 mass protests, strikes, and mourning

[119] Mehdi Moslem, *Factional Politics in Post-Khomeini Iran* (Syracuse University Press, 2002), 29–45.

[120] Geoffrey Robertson, *The Massacre of Political Prisoners in Iran, 1988* (Abdorrahman Boroumand Foundation, 2011); Azadeh Moaveni, *Lipstick Jihad: A Memoir of Growing Up Iranian in America and American in Iran* (PublicAffairs, 2005).

[121] Augustus Richard Norton, *Hezbollah: A Short History* (Princeton University Press, 2007), 15–32.

RTVC Stage	Iranian Revolution Summary
Revolution	rituals escalate; Black Friday massacre a trigger.
Civil War (Optional Branch)	Avoided; military defection and regime collapse occured without protracted armed conflict.
Revolutionary Victory	Provisional Government takes power
Coup d'État (Optional Branch)	Khomeini returns, purges rivals, establishes Islamic Republic through referendum and new constitution. Revolutionary Courts used.
Tyranny	Khomeini consolidates power as Supreme Leader; Basij militia and IRGC enforce repression; 1988 prison massacres and ideological indoctrination.
Export of Terror	Formation of Hezbollah; regional support for Shi'a militant groups (e.g., in Iraq, Syria, Yemen).
Democratic Transition	Short-lived attempt by Shahpour Bakhtiar fails; no sustained democratic off-ramp.

The RTVC model provides a structured framework for understanding how terrorist movements can transition into governing regimes. While not every case aligns perfectly with the model, it highlights key indicators that analysts and policymakers can use to assess, and potentially intervene in, emerging revolutionary movements. Historical examples such as Russia and Iran demonstrate that, although not all groups progress through every stage, many revolutionary movements follow recognizable patterns. This insight enables more effective predictive analysis and informs counterterrorism strategies. Ultimately, countering revolutionary terror requires prevailing in the ideological arena and disrupting groups before they advance through these critical stages.

Originally written in July 2010, my article *Terrorist to Tyrant,* predicted mass civil unrest in the Middle East as the next stage in a Sunni Islamic revolutionary Terror Value Chain.

> *What, then, can we expect to see in the development of international Islamic revolutionary terrorism? Although it remains a threat, the core group of al Qaeda has been heavily damaged by the U.S. reaction to the 9/11 attacks. Using the Revolutionary Value Chain, this places us in the third stage: Mass Civil Unrest and Revolution. Of greater concern than individual terror attacks is the wider movement inspired and motivated by jihadist ideology.*[122]

It is therefore that we now turn to the Arab Spring.

---

[122] Myers, "Terrorist to Tyrant," 17.

CHAPTER 2

JUSTICE AND JIHAD:
The Arab Spring

Ten years on, the hopes awakened by the protests have vanished — but the underlying condition that drove the unrest are as acute as ever.[123]

Following the fall of the Ottoman Empire, the Muslim world had been, for lack of a better term, dormant. In his *The Clash of Civilizations*, Samuel Huntington describes an Islamic Resurgence beginning in the 1980s as being of similar magnitude and historical importance as the Russian Revolution or the Protestant Reformation.[124] This Muslim awakening was a widespread generalized social movement, largely rejecting Westernization and seeking instead to reclaim Islam's former position of influence in the world. In 2007, Osama Bin Laden referred to the "humiliation and disgrace that Islam suffered more than eighty years ago," perhaps meaning the 1918 occupation of Istanbul and the partition of Ottoman territories by the Allied victors.[125]

Preceded by the campaigns of terror perpetrated by the Egyptian Islamic Jihad and Al Qaeda, the revolutionary wave known as the Arab Spring has been followed by bloody conflicts throughout the Middle East. The Arab Spring itself, which began in in December 2010, toppled regimes in Tunisia, Egypt, Yemen, and Libya, and plunged Syria into a devastating civil war that lasted more than a decade. The aftermath has been marked by bloody conflict,

---

123 Liz Sly, "Lost Decade: The unfinished business of the Arab Spring," *Washington Post*, January 24 2021, https://www.washingtonpost.com/world/interactive/2021/arab-spring-10-year-anniversary-lostdecade/.

124 Samuel Huntington *The Clash of Civilizations and the Remaking of World Order* (New York: Simon & Schuster, 1996), 109.

125 Bernard Lewis, "The Revolt of Islam: When did the conflict with the West begin, and how could it end?" *The New Yorker*, November 19, 2001.

displacement, and the rise and fall of new political actors. This chapter explores the Arab Spring through the lens of the Revolutionary Terrorism Value Chain (RTVC) framework, offering a comprehensive comparative analysis of how four countries, Tunisia, Libya, Egypt, and Syria, navigated the revolutionary process.

The demands for justice and change voiced by millions of Arabs were rooted in legitimate grievances: authoritarian rule, economic hardship, and lack of political freedom. Hope surged that the Arab Spring would usher in a new era of popular sovereignty. Yet, as with many revolutions, the promise of democracy soon collided with the ambitions of those seeking power.

In Egypt, the initial euphoria gave way to military intervention and the reassertion of authoritarian control, while in Syria, the revolution devolved into a protracted civil war, only recently concluded with the fall of the Assad regime and the emergence of a transitional government rooted in Islamic ideology. In Libya, the overthrow of Muammar Gaddafi led to state collapse and ongoing fragmentation among competing militias. Only Tunisia managed a fragile democratic transition, though even there, authoritarian pressures persist. The outcomes remain uncertain, but the original drivers of unrest, calls for justice, dignity, and self-determination, continue to shape the region's turbulent political landscape.

The Arab Spring provides particularly rich terrain for testing the RTVC framework because the four countries examined here, Tunisia, Libya, Egypt, and Syria, all experienced revolutionary upheaval within the same time period, driven by similar underlying grievances, yet produced dramatically different outcomes. This natural experiment allows us to isolate the variables that determine whether revolutions lead to democracy, renewed tyranny, or state collapse.

Jihad, Just War, and the Theology of Legitimate Rebellion

The chapter's title, Justice and Jihad, requires examination of how Islamic legal and theological traditions address revolution, rebellion, and the use of force. The Arab Spring's religious dimension cannot be understood without grasping the centuries-old debate within Islamic jurisprudence about when, if ever, Muslims may legitimately rebel against an unjust ruler.

Classical Islamic Just War Theory[126]

Islamic legal scholars developed a sophisticated jurisprudence of war and rebellion from the Qur'an, hadith, sīra, and the accumulated reasoning of the legal schools. Modern scholars often compare this tradition to Western just-war

---

[126] Andrew F. March, "Ambivalent Universalism? Jus ad Bellum in Modern Islamic Legal Discourse," *European Journal of International Law* 24, no. 1 (2013): 367–389.

reasoning, though the analogy is imperfect because Islamic law is not a single codified system but a juristic tradition of argument, precedent, and interpretive authority. Classical jurists distinguished defensive jihad from expansionary or initiative jihad, while modern scholars have debated whether the latter reflected a permanent norm, a historically contingent feature of premodern international politics, or a category now superseded by treaty-based relations among sovereign states. The Arab Spring reopened a related but distinct question: whether public protest or rebellion against Muslim rulers constitutes prohibited *khurūj* and fitna, or whether nonviolent resistance to tyranny can be justified within the higher objectives of Islamic law.

The Problem of Khurūj: Rebellion Against Muslim Rulers[127]

The Arab Spring's theological complexity centers on khurūj—armed rebellion against a Muslim ruler. Classical Islamic jurisprudence strongly discouraged rebellion even against unjust rulers, based on the priority of social order and prevention of fitna (civil strife). The dominant Sunni position, articulated by scholars like Imam al-Ghazali, held that an unjust ruler is better than the chaos of civil war.[128] The tyrant must be endured unless his rule becomes so oppressive that greater harm comes from obedience than rebellion.

This conservative stance rested on several foundations: the Quranic injunction to obey authority (Surah Al-Nisa 59), historical trauma from the first Islamic civil war following Caliph Uthman's assassination, and pragmatic calculation that rebellion rarely succeeds and typically produces worse tyranny.[129]

However, Classical and later Sunni jurists identified narrow conditions under which rebellion could become permissible, or in some readings obligatory: manifest disbelief (*kufr bawāḥ*) by the ruler, clear proof, sufficient capacity or reasonable prospect of success, and enough organization or *manāʿa* among the rebels to prevent rebellion from collapsing into mere fitna or criminal disorder.[130]

---

127 Muhammad Al-Atawneh, "Khurūj in Contemporary Islamic Thought: The Case of the Arab Spring," *Ilahiyat Studies* 7, no. 1 (2016): 27–52.

128 Albert Hourani, *A History of the Arab Peoples* (Cambridge, MA: Belknap Press of Harvard University Press, 1991), 144.

129 Al-Nisa 59: O you who believe, obey Allah and obey the Messenger and those in authority from among you; then if you quarrel about anything, refer it to Allah and the Messenger, if you believe in Allah and the Last Day. This is best and more suitable to (achieve) the end.

130 Sadia Tabassum, "Combatants, Not Bandits: The Status of Rebels in Islamic Law," *International Review of the Red Cross* 93, no. 881 (2011): 259–80.

Radical Reinterpretation: Qutb and Revolutionary Islam

Sayyid Qutb's innovation was to radically lower the threshold for takfir (declaring someone an unbeliever) and thus for legitimate rebellion. His key claims: (1) any society not governed by sharia exists in pre-Islamic ignorance (jahiliyya), (2) leaders implementing secular law commit apostasy and forfeit legitimacy, (3) a righteous vanguard must wage jihad to overthrow such systems, even if success is uncertain, and (4) the struggle continues until the Islamic state is established globally.

This represented a revolutionary break with classical Sunni jurisprudence. Where traditional scholars required extraordinary evidence before declaring a Muslim an apostate and permitted rebellion only in extreme circumstances, Qutb essentially declared all existing Muslim governments illegitimate, transforming khurūj from a rare emergency exception into a permanent revolutionary duty.

Moderate Islamic Responses

Not all Islamists embraced Qutb's radicalism. The mainstream Muslim Brotherhood (at least rhetorically) rejected violence in favor of gradual Islamization through education, social services, and democratic participation. The Arab Spring tested these competing visions, in Tunisia and Egypt, Islamists initially pursued democratic participation; in Libya and Syria, state collapse created space for armed jihad. The failure of moderate Islamism in Egypt, where democratically elected Mohamed Morsi was overthrown by the military, seemed to vindicate Qutb's argument that the system cannot be reformed from within, pushing some toward radicalization.

The interplay between competing Islamic legal theories of rebellion, Western intervention, authoritarian repression, and popular aspirations for justice created the volatile mixture that defined the Arab Spring. Understanding "Justice and Jihad" requires recognizing that these were not merely political struggles but theological contests over the meaning of legitimate Islamic governance, the criteria for just war, and the permissibility of rebellion against Muslim rulers. We now turn to systematic application of the RTVC model to four Arab Spring cases.

COMPARATIVE CASE STUDIES

The Arab Spring was a pan-Arab, transnational phenomenon sharing much of the foundational elements. The Islamism of Hassan al-Banna and Said Qutb lay the ideological foundation for al Qaeda's global terror organization. Al Qaeda's goal was to overthrow the existing order in the Middle East. To do so it had to undermine the support provided to governments in the region. Therefore, the regime was targeted by the Islamists, which provoked a massive reaction. The American invasion of Afghanistan and especially of Iraq created much of the unrest which led to the Arab Spring uprisings. Although the Arab

Spring was sparked in Tunisia, the pan-Arab, transnational Arab revolution originated with al-Banna and Qutb in Egypt.

Egypt: Revolution, Counterrevolution, and the Return of Tyranny

Egypt's revolutionary trajectory demonstrates how competing ideologies, military intervention, and failed consolidation can lead a promising democratic opening to collapse into renewed authoritarianism. The RTVC framework reveals Egypt's path through revolution, brief democratic experiment, military coup, and return to tyranny.

Historical Context: Islamists and Nationalists

To understand Egypt's Arab Spring, we must first examine the decades-long contest between Islamism and Arab Nationalism that shaped Egyptian politics.

Since the mid-1950s, Arab politics have been polarized—a struggle between Islamists and nationalists. There is no third wave or viable alternative to compete with the ideological hegemony of the Islamists on one hand and the nationalist-military alliance on the other. While the Arab Spring uprisings briefly raised hopes that such an alternative would emerge, those hopes were dashed by collusion between the two sworn enemies. While there were Marxist voices in the Arab Spring protests, they were marginal. The real contest, both before and after the uprisings, was between Arab nationalism (represented by the military and Nasserist ideology) and political Islam (led by the Muslim Brotherhood and its offshoots).[131]

The British occupied Egypt in 1882 as the Ottoman Empire began to crumble in the decades prior to World War I. Following the war, the French and British divided the spoils of the empire including Syria. Huntington's Sunni Islamic Resurgence had its roots at this time; in 1928 Hassan al-Banna founded the Muslim Brotherhood in Egypt. The Brotherhood "stressed the necessity of a struggle not only to liberate Egypt but the whole of the Islamic homeland." Rising political violence in Egypt led to the Brotherhood being outlawed in 1948 and al-Banna's assassination the following year. Thus, of necessity, it developed into a clandestine organization.[132]

Muhammad Ghazzali, who later left the ranks of the Brotherhood, developed the idea that the West was continuing the Crusades of the Middle Ages. He declared a need to uphold cultural values in the face of Western depravity seeping into Muslim society. More extreme members of the

---

[131] Fawaz A. Gerges, *Making the Arab World* (Princeton: Princeton University Press, 2018), 393, Kindle.

[132] Hamid Enayat, *Modern Islamic Political Thought*, (New York: I.B. Tauris & Co Ltd., 2005) 85.

Brotherhood, such as Sayyid Qutb, became disillusioned with the non-violent stance the Brotherhood adopted in the face of repression and broke away as well. Qutb felt Islamic society could only come about through the destruction and replacement of the existing secular power.

Qutb eschewed gradualist political engagement and social mobilization in favor of nourishing a subversive vanguard that would spearhead the institution of a new Islamist utopia. Qutb and his followers dismissed politics and activism as ineffectual and instead prioritized armed struggle and subversion against Nasser, whom they labeled the taghut (tyrant).[133]

According to Qutb, the world was in a state of *jahiliyyah*, or ignorance before Islam's enlightenment. He was influenced in his thinking by the Imam Sayyid Abul A'la Mawdudi, who described the "modern *jahiliyyah*" as modern secularism. Mawdudi founded Jamaat-e-Islami in Pakistan to create an Islamic state free of secularism. Qutb advocated violent opposition to secularism, which he believed led to the decline of morals; according to Qutb, men must choose either material comforts or Islam. He also called on Muslims to physically fight to overcome secularism. His denunciations of Muslim leaders, both secular and religious, ultimately resulted in his execution in 1966. The ideas he championed permitted jihadists to rationalize terrorism as a just defense of the faith. Qutb's manifesto, *Milestones*, is comparable in its impact to Rousseau's *Social Contract* and Lenin's *What Is to Be Done?* with similarly bloody consequences. In Gaza, Muslim Brothers influenced by Qutb founded the Islamic Jihad which would become known as Hamas.[134]

The Nasserist Era

The Muslim Brotherhood and the Free Officers, a clandestine nationalist military organization formally led by Mohammed Naguib, had initially joined ranks to overthrow the British-backed monarchy in 1952. When the Free Officers, staged a coup overthrowing the British-backed Egyptian monarchy the Muslim Brotherhood expected to be part of the new ruling coalition; however, they were soon sidelined by the Free Officers and largely went underground.

The Free Officers acted systematically to dismantle the multiparty system and the power structure of the old regime, through measures such as the passage of Qanoun Tanzim al-Ahzab (the Law for the Organization of Political Parties), formalized by the cabinet on September 9, 1952. Although they had promised that the party law would be judicially executed under the supervision of the State Council, they never respected their pledge. In his autobiography, Naguib laments that decision by the State Council as marking a descent toward

---

[133] Gerges, *Making the Arab World*, 242.
[134] Ibid., 7.

political authoritarianism, and he acknowledges that the law was designed to disinherit the old social classes and liquidate political rivals, particularly the nationalist Wafd, which remained the biggest and most popular political party until 1952.

One of the officers, Gamal Nasser, purged his own organization, sidelining Naguib, and then turned on the Muslim Brotherhood; thousands were rounded up and imprisoned. The Free Officers' crackdown on the Brotherhood morphed into an armed clash between the Arab nationalist project represented by the Nasserist state and a radical Islamists led by Qutb. A paramilitary wing, Al-Tanzim al-Sirri (the Secret Organization), was established by Brotherhood members after Nasser's suppression of the movement in 1954.[135] Qutb played a pedagogical-ideological role and also provided practical guidance and supervision to a relatively large number of dedicated young religious activists both outside and inside Egypt's prisons.

Sadat's Opening and Assassination

The Brotherhood reappeared under Nassar's successor, Anwar Sadat, who was initially friendly to them. It was a reciprocal relationship in which Sadat's interests and those of the Islamists dovetailed. Sadat treated Islamists with "a benevolence that was well reciprocated, as the Islamists 'purged' the universities of anything that smelled of communism."[136]

But the relationship did not last. Sadat was assassinated by the Egyptian Islamic Jihad in 1981. "Sadat thought he could rely on us because in his eyes we owed our existence to him and we would not dare challenge his regime and rise up against him; though when we subsequently discovered his lies, we killed him," recalled the Islamist attorney Montasser al-Zayat.[137]

Mubarak's Repression

Sadat's successor, Hosni Mubarak, took a hardline against all Islamists, driving the Brotherhood underground once again. The rise of militant Islamist groups such as Egyptian Islamic Jihad and the Islamic Group (Gama'a Islamiyya), whose ranks included future Al Qaeda leader Ayman al-Zawahiri, made even the ultraconservative old guard within the Brotherhood appear moderate by comparison.[138]

The merger of Zawahiri's al-Jihad organization with al-Qaeda added clandestine expertise of the Egyptians to the financial resources of Bin Laden.

---

135 Gerges, Fawaz A.. Making the Arab World (p. 372). Princeton University Press. Kindle Edition.

136 Ibid. 325.

137 Ibid, 323-324.

138 Ibid, 369.

Al Qaeda was then able to create a global underground organization based on a network of cells. These cells soon began attacking Western interests and eventually elicited the desired response: the U.S. invasion and occupation of Muslim lands. Al Zawahiri and Bin Laden wanted this response, which they hoped would motivate the Muslim population to overthrow the existing powers and reestablish the Islamic Caliphate in the Middle East.

The global jihadist movement, spearheaded by al-Qaeda, has propagated a narrative that highlights the action, excitement, and camaraderie of joining the fight for the caliphate. And ISIS excelled at tailoring its messages to myriad demographics and lowering the barriers to entry for those not well steeped in Islamic theology, culture, or practice.[139]

Egypt's Arab Spring: Systematic RTVC Analysis

Ideological Development

The Muslim Brotherhood diagnosed the Mubarak regime as corrupt and anti-Islamic, envisioning a just society governed by sharia law. There were also secular liberals and leftists demanding social justice and democracy. Both camps justified extraordinary means, especially after decades of failed reform.

While the uprising united Egyptians against the regime in January 2011, the revolutionary coalition lacked a cohesive, unifying ideology. Competing visions, secular, Islamist, and nationalist, quickly surfaced, leading to fragmentation and mistrust among key actors once Mubarak fell.

The Vanguard

The Muslim Brotherhood, wary of the many cycles of repression it had suffered, initially sat out the protests, joining only when Mubarak's fall seemed inevitable. This late entry would later undermine their legitimacy among revolutionary youth. Nevertheless, the clandestine networks of the Muslim Brotherhood and radical offshoots like Egyptian Islamic Jihad were organized and ready to seize the opportunity when it presented itself.

Youth activists advised by the Serbian organization CANVAS were able to mobilize a mass uprising coordinated using social media. The movement however, lacked a disciplined, unified vanguard capable of guiding the transition.[140] The Muslim Brotherhood, on the other hand, had the organizational capacity to dominate the post-Mubarak political landscape. With millions of members, extensive social service networks, and decades of political experience (albeit underground), the Brotherhood was the only opposition force with nationwide organization. This organizational advantage, however,

---

139. J. M. Berger and Jessica Stern, *ISIS: The State of Terror* (New York: Ecco Press, 2015), 83–84.

140 Tina Rosenberg, "Revolution U: What Egypt Learned from the Students Who Overthrew Milosevic," *Foreign Policy*, February 16, 2011

came with ideological baggage and a history of confrontation with the military that would prove fatal.

Mass Civil Unrest and Revolution

The American response to 9/11, the so-called Global War on Terror (GWOT), along with the push for Western-style democracy, increased existing resentment among the Arab population. Adding to economic stress, the repression of Islamists by regional secular regimes heightened dissatisfaction that finally exploded in the Arab Spring. The catalyst for the uprising was the self-immolation of Mohamed Bouazizi in Tunisia in December 2010. Unrest in Tunisia quickly spread across the region.

The Revolution in Egypt began on January 25, 2011, bringing about President Mubarak's resignation on February 11. Applauding the uprisings, many in the West hoped democracy would break out. The Marxist theorist Emmanuel Wallerstein went so far as to liken the Arab Spring to the revolutionary wave that swept the world in 1968. But the organizers of the protests that forced Mubarak from power did not seem to have a plan for the aftermath. To gain the broadest support possible, the organizers left their agenda vague; "Democracy went unmentioned."[141]

Mass protests succeeded in toppling Mubarak and briefly opening the political space. However, the lack of coordination and unity among revolutionary forces allowed the military (operating as the Supreme Council of the Armed Forces, or SCAF) and remnants of the old regime to maintain leverage and shape the transition process. As partisanship deepened, the revolutionary coalition fractured, and the ability to sustain pressure on the old power structures waned.

Revolutionary Victory and Consolidation of Power

In 2011, the Muslim Brotherhood gained power through democratic elections but could not retain it. President Mohamed Morsi's attempt to centralize power and sideline rivals provoked mass protests. The SCAF retained significant power and influence throughout the transition, positioning itself as the ultimate arbiter. The Brotherhood divorced itself from coalition allies but could not consolidate its newfound position without the Army's support. The military, led by General Abdel Fattah al-Sisi, staged a coup in July 2013, purging the Brotherhood and restoring authoritarian rule.

Massive street protests on the anniversary of his election, which spurred the military's deposing of Morsi, were driven in part by perceptions that the Brotherhood was trying to concentrate power in its own hands. Morsi unilaterally amended the constitutional declaration to shield his decisions from

141 Immanuel Wallerstein, "The Contradictions of the Arab Spring," *Al Jazeera*, November 14, 2011.

judicial review; his opponents painted it as an unjustified power grab.

Morsi did not purge the bureaucracy of old regime loyalists, perhaps only because he did not have the power to do so. While it is unlikely Morsi will return to power, the unrest in Egypt is probably unfinished. It is worth noting 12 years passed between the failed 1905 revolution and the successful 1917 revolution in Russia, and the Muslim Brotherhood has survived since 1928.

Tyranny

After the 2013 coup, the military outlawed the Muslim Brotherhood and brutally repressed its members. Secularists, who had hoped to dominate electoral politics and reset the country's democratic transition after the military coup, instead faced a military intent on establishing a more comprehensive dictatorship than the one that had prevailed during Hosni Mubarak's tenure.

Sisi's regime institutionalized repression, criminalized opposition, and built a cult of personality. The revolutionary promise of justice and democracy was replaced by a narrative of order and security, returning Egypt to authoritarian guardianship that echoed Egypt's historical patterns of state dominance.

From the Muslim Brotherhood's perspective, al-Sisi is a tyrant waging a campaign of repression against Islam. Perhaps, however, fearing the violence seen in Syria and Libya, or that his fall will lead to violence akin to the French Revolution's Terror, a large portion of the population remains loyal to al-Sisi.

Although the nationalists, led by the military, hold the advantage for now, they have not converted their victory over the Islamists into enduring political and social capital. Theirs may be a hollow victory. They must deliver prosperity and justice to the people. "Dictators have prevailed, mainly through coercion," said Lina Khatib, who heads the Middle East and North Africa program at London-based Chatham House. "However, coercion seeds further grievances that will ultimately force citizens to seek political change." Worse instability and violence could result if the government cannot address systemic grievances.

Export of Terror and Expansion

The crackdown on the Muslim Brotherhood and other opposition groups contributed to the radicalization of some Islamists, fueling violence in Sinai and beyond. An ISIS affiliate established itself in the Sinai Peninsula, conducting attacks against Egyptian security forces and Coptic Christians. However, the Sisi regime itself focused on consolidating internal control rather than exporting revolution because his coup was reactionary rather than revolutionary.

Off-Ramp to Democratic Transition

There was hope in the West that Egypt would follow Tunisia in transitioning to democracy. In Tunisia, the Muslim Brotherhood (operating as the Ennahda party) shed some of its Islamist exclusivism and was able to cooperate with secular political parties to form an effective and moderate government. In Egypt, the Muslim Brotherhood also won elections, gaining the presidency for Mohammed Morsi.

However, the Muslim Brotherhood's failure to adopt a more conciliatory stance and a more open mind towards its political rivals was one of the most notable blunders made by the leadership during and after the January revolution. It fed into the fears of the Egyptian public and consequently allowed the military to carry out a counterrevolution and remove the group from power after it had won the 2011 presidential elections. One of the major criticisms leveled against Mohamed Morsi's presidency was his growing autocracy, secrecy, and opaque decision making.

Additionally, Morsi's inability to improve the economy increased public discontent. The military security apparatus remained intact and, with the support of much of the public, removed Morsi and the Brotherhood from power.

Egypt failed to take the RTVC's democratic "off-ramp." The military's dominance, the Brotherhood's political missteps, and the absence of strong, independent institutions prevented a negotiated transition.

Civil War (Optional Branch)

Civil war was averted, but only through overwhelming military force and repression rather than reconciliation or reform. The Sinai insurgency represents a low-level conflict, but Egypt proper avoided the state collapse seen in Libya or Syria.

Coup d'état (Optional Branch)

The military coup in Egypt ended the democratic experiment and ignited a new phase of violent conflict between the state and Islamist groups, which saw the military's generalized repression of any opposition.

Holger Albrecht compares the collective action by the Egyptian military with the actions of the militaries in Syria, Tunisia, and Yemen.[142] In Syria, large numbers of rank-and-file combatants took individual action and deserted. A large part of the military collectively mutinied in Yemen, and in Tunisia, the Chief of Staff took individual action, refusing orders to suppress the uprising, in effect defecting from the regime.

In the context of the Arab Spring, the Egyptian military conducted a coup against Morsi's elected government. But rather than a radical coup, in Egypt it was counterrevolutionary; the June 30, 2013, Tamarod (Rebellion) movement targeted Morsi, and in effect replaced the pre-Arab Spring leader Hosni

---

142 Holger Albrecht, "Military Insubordination in Popular Mass Uprisings," *Political Science Quarterly* 134, no. 2 (Summer 2019): 303–328.

Mubarak with Al-Sisi.

The 2013 coup, led by Sisi, exemplified a regime insider's seizure of power during the consolidation phase, decisively ending the democratic experiment.

Summary

According to the RTVC framework, Egypt's return to tyranny was driven by a fragmented revolutionary ideology, the absence of a unified vanguard, the military's enduring power, failed coalition-building, and the rapid reassertion of authoritarian mechanisms. The revolutionaries lost leverage over the old regime, enabling state institutions, especially the military, to absorb the revolutionary shock, suppress dissent, and restore authoritarian rule.

Tunisia: The Jasmine Revolution—A Fragile Democratic Success

The RTVC framework helps explain Tunisia's comparatively successful democratic transition by highlighting how the country navigated each stage of revolutionary transformation differently from its regional peers.

Ideological Development

Tunisia's revolution began with a clear articulation of grievances, corruption, unemployment, repression, and lack of political freedoms, crystallized by Mohamed Bouazizi's self-immolation in December 2010. The movement recognized the failures of the Ben Ali regime and envisioned a pluralistic, democratic society that would restore dignity and justice. This narrative unified diverse segments of society and justified extraordinary mobilization, but crucially, it did not devolve into a call for violent insurrection or rigid ideological purity.

The Tunisian revolutionary ideology remained inclusive rather than exclusionary, democratic rather than theocratic, and reformist rather than nihilistic. This ideological moderation created space for compromise and prevented the factional splintering that plagued other Arab Spring movements.

The Vanguard

Unlike in Libya or Egypt, Tunisia's vanguard was not a clandestine militant group but a coalition of civil society organizations, labor unions (notably the UGTT, Tunisian General Labour Union), the Tunisian branch of the Muslim Brotherhood (Ennahda), and professional associations.[143] These actors operationalized the revolutionary ideals through coordinated, largely nonviolent action and mediation, rather than through terror or armed struggle.

Their legitimacy and organizational strength enabled them to channel mass discontent into structured political dialogue and negotiation. The vanguard's

143 Sebastian Benn, "The Power of Labor: Tunisia's Trade Union and the Arab Spring," *theSquare Centre*, August 1, 2024.

civic rather than military character proved decisive in preventing the revolution from militarizing and maintaining pressure on the regime without destroying state institutions.

Mass Civil Unrest and Revolution

The revolution quickly moved from fringe dissent to mass mobilization, with broad participation from youth, workers, and the middle class. Overt political parties and coalitions formed, and mass protest actions, strikes, marches, and sit-ins, pressured the regime. The military's decision to remain neutral and the police's eventual sympathy for protesters prevented large-scale bloodshed and helped avoid a spiral into civil war.

The breadth of participation, crossing class, regional, and ideological lines, made the movement difficult to repress or co-opt. When Ben Ali's security forces killed protesters, it only intensified public outrage rather than quelling dissent.

Revolutionary Victory and Consolidation of Power

Ben Ali fled on January 14, 2011, leaving a power vacuum. A moderate provisional government formed, but the real consolidation came through a negotiated process: the National Dialogue Quartet (comprising the UGTT, the Tunisian Order of Lawyers, the Tunisian Human Rights League, and the Tunisian Confederation of Industry, Trade and Handicrafts) brokered compromises between Islamists (Ennahda party) and secularists, leading to a new constitution and free elections.[144]

Crucially, purges and state terror were largely avoided. The process was inclusive and consensus-driven, rather than dominated by a single, ruthless faction. Ennahda, despite winning elections, agreed to share power and eventually stepped back, demonstrating commitment to democratic norms over raw power retention. And, perhaps, Tunisia was spared revolutionary violence because many of its most radical citizens were not in the country having left to participate in jihad elsewhere.

Tyranny

Tunisia largely avoided the descent into tyranny typical of many revolutions. While there were moments of political violence and crackdowns on Salafi extremists, the state did not institutionalize widespread repression or cults of personality. The absence of a militant vanguard seeking to monopolize power prevented the revolutionary state from becoming a terror apparatus.

Recent years have seen democratic backsliding under President Kais Saied, who dissolved parliament in 2021 and centralized power. Yet even this

---

144 Counter Extremism Project, "Muslim Brotherhood in Tunisia," accessed May 22, 2026, https://www.counterextremism.com/content/muslim-brotherhood-tunisia.

authoritarian drift has not reached the levels of repression seen in Egypt or Assad's Syria, and significant civil society resistance continues.

Export of Terror and Expansion

Tunisia did not attempt to export its revolution militarily or ideologically. Instead, it focused on internal reforms and building democratic institutions. However, the democratic model inspired activists elsewhere in the region during the early Arab Spring period.

Paradoxically, Tunisia also became a major source of foreign fighters joining jihadist groups, especially ISIS, with thousands of Tunisians traveling to fight in Libya and Syria.[145] This phenomenon reflects the incomplete nature of Tunisia's transition, economic grievances remained unresolved, creating recruitment opportunities for extremists.

Off-Ramp to Democratic Transition

Tunisia exemplifies the RTVC's rare "off-ramp." The military remained neutral rather than seizing power, civil society was robust and well-organized, and moderate leadership prevailed on all sides. The 2014 Constitution enshrined decentralization, pluralism, and participatory governance, further embedding democratic norms.[146]

The National Dialogue Quartet's mediation prevented the winner-take-all dynamics that plagued Egypt, where the Muslim Brotherhood's electoral victory led to polarization and military coup. Instead, Tunisia developed power-sharing arrangements and coalition governments that, while messy, prevented descent into authoritarianism or civil war.

Civil War (Optional Branch)

Tunisia avoided civil war thanks to institutional compromise, the absence of armed insurgency, and the ability of civil society to mediate between competing factions. The military's neutrality was critical, had it intervened on either side, civil conflict might have erupted.

Coup d'état (Optional Branch)

No coup disrupted the transition, in stark contrast to Egypt. The Tunisian military's professionalism and relatively small size (it had been deliberately kept weak by authoritarian rulers) meant it lacked both the capacity and incentive to seize power.

Summary

---

[145] Aaron Y. Zelin, "Tunisians of the Iraq Jihad and How That Set the Stage for the Syrian Jihad," *The Washington Institute for Near East Policy*, January 29, 2021.

[146] Mohamed Kerrou, "Tunisia's Historic Step Toward Democracy," *Carnegie Endowment for International Peace*, April 22, 2014; *The Constitution of the Tunisian Republic*, adopted January 26, 2014, arts. 14, 18, 35, 139

The RTVC framework shows that Tunisia's success stemmed from a unifying, inclusive ideology; a nonviolent and legitimate vanguard rooted in civil society; mass mobilization that pressured but did not destroy state institutions; and, most importantly, a willingness to compromise and negotiate among diverse actors. This enabled Tunisia to achieve revolutionary victory without descending into tyranny, civil war, or military rule, setting it apart from its neighbors.

Yet Tunisia's democratic transition remains fragile. Economic challenges persist, with high unemployment especially among youth. The fall of the Assad regime has led to fears that Tunisian jihadists will return and bring more violence to the country.[147] President Saied's 2021 power grab demonstrates the continuing tension between democratic aspirations and authoritarian temptations. The current Constitution adopted in 2022 moved Tunisia away from the semi-parliamentary compromise of 2014 and toward a much stronger presidency.[148] Whether Tunisia can consolidate its democratic gains remains an open question.

Libya: Revolution, Fragmentation, and Enduring Chaos

Libya's ongoing fragmentation is deeply rooted in the absence of a cohesive revolutionary ideology and the weakness of its vanguard during and after the 2011 uprising. The RTVC framework illuminates how Libya's revolution failed at nearly every critical juncture.

Ideological Development

The revolutionary movement was driven more by shared opposition to Muammar Gaddafi than by a clear, positive program for Libya's future. This absence of a compelling, shared ideology meant there was no common framework for governance or national identity to bind the various factions together once Gaddafi was defeated.

Calls for extraordinary means, including armed rebellion, were quickly justified by Gaddafi's brutal crackdown on peaceful protesters in February 2011. However, the movement never developed beyond "anti-Gaddafi" to articulate a coherent vision of what Libya should become. Regional identities (Cyrenaica vs. Tripolitania), tribal affiliations, and Islamist vs. secular divisions all competed without any unifying revolutionary narrative to bridge these cleavages.

---

[147] Fatima Badwi, "Tunisia-Syria-Tunisia, The Human Tide of Jihadists Is Turning Again," *Worldcrunch*, January 27, 2025, updated May 24, 2025, translated and adapted by Elias Kassem, originally published by *Daraj*.

[148] *The Constitution of the Republic of Tunisia, 2022*, adopted July 25, 2022, published August 18, 2022, WIPO Lex, https://www.wipo.int/wipolex/en/legislation/details/21853.

The Vanguard

Libya's revolutionary vanguard was fragmented from the outset. Instead of a disciplined, ideologically driven core, the uprising was led by a loose coalition of militias, local leaders, tribal elders, and defected military officers, each with their own interests and limited coordination.

The National Transitional Council (NTC), established in Benghazi in February 2011, attempted to provide political leadership but never achieved operational control over the various armed groups fighting Gaddafi.[149] Each militia maintained its own command structure, funding sources, and territorial ambitions. This fragmentation proved fatal to post-revolutionary state-building.

Mass Civil Unrest and Revolution

Protests in Benghazi and Tripoli in February 2011 escalated rapidly into mass rebellion. As the regime cracked down with live ammunition and air strikes, militias proliferated across eastern Libya. The opposition's political wing, the NTC, sought international recognition and received it, bringing NATO intervention in March 2011.

NATO's air campaign, justified by humanitarian concerns and authorized by UN Security Council Resolution 1973, shifted the military balance decisively.[150] Without NATO air power, the rebellion might have been crushed. However, NATO's intervention contributed to Libya's fragmentation by enabling dozens of militias to survive and succeed without unifying under central command.

Revolutionary Victory and Consolidation of Power

Gaddafi was overthrown and killed by rebel forces in October 2011. However, the collapse of central authority led to a fractured landscape. Rival militias and tribal factions vied for power, and no single group could consolidate control. Attempts at purges and disarmament failed; assassinations of former regime officials and rival militia leaders became commonplace.

The NTC proved unable to establish a monopoly on violence, the fundamental requirement for any state. Militias refused to disarm, maintaining their own fiefdoms. Benghazi, Misrata, Zintan, and other cities became semi-autonomous zones controlled by local armed groups.

Tyranny

Instead of a new centralized tyranny, Libya descended into statelessness, multiple local strongmen and jihadist factions imposed their own forms of violence and repression, but no centralized regime emerged to exercise

---

[149] Atlantic Council, *Libya's Transition* (Washington, DC: Atlantic Council, November 15, 2012), 2, 5.

[150] NATO, "NATO and Libya (February–October 2011)," updated November 8, 2015.

nationwide terror. This diffusion of violence, paradoxically, made Libya's situation worse than traditional tyranny, as no single actor could be held accountable or negotiated with.

Ansar al-Sharia, ISIS affiliates, and other extremist groups established control in parts of Benghazi and Derna, implementing harsh versions of Sharia law.[151] The 2012 attack on the U.S. consulate in Benghazi exemplified the chaos, as no central authority could prevent or respond to such incidents.

Export of Terror and Expansion

Libya became a hub for arms trafficking, jihadist training, and foreign-fighter facilitation, fueling conflicts in Mali, Syria, the Sinai, and beyond. Gaddafi's massive weapons stockpiles were looted after the regime's collapse and moved across the Sahel and North Africa; while the strongest evidence concerns Mali, Niger, Tunisia, Sinai, Gaza, and Syria, smaller or less fully confirmed traces of Libyan-origin weapons also appeared as far afield as Nigeria and the Central African Republic.[152] The state's collapse enabled the export of violence and weapons, not revolutionary ideology per se, but instability itself.

Off-Ramp to Democratic Transition

Attempts at democratic transition failed due to weak institutions, lack of national identity, and the dominance of armed groups. Elections were held in 2012 and 2014, but winning parties couldn't govern without military backing, which they lacked. The off-ramp was never viable because no political force commanded sufficient legitimacy or power to implement democratic norms.

Civil War (Optional Branch)

Libya's revolution quickly morphed into a multi-sided civil war. By 2014, two rival centers of authority claimed legitimacy: a Tripoli-based government backed by the Libya Dawn coalition, which included Islamist and Misratan armed groups, and a Tobruk-based House of Representatives, internationally recognized at the time and aligned with General Khalifa Haftar's Libyan National Army. ISIS exploited the resulting vacuum, establishing its most important Libyan stronghold in Sirte between 2015 and 2016.[153]

---

151 Mapping Militants Project, "Ansar al-Shariah (Libya)," last modified July 2018, https://mappingmilitants.org/node/327.

152 Nicholas Marsh, "Brothers Came Back with Weapons: The Effects of Arms Proliferation from Libya," *PRISM* 6, no. 4 (2017): 79–96; United Nations Security Council, *Final Report of the Panel of Experts Established Pursuant to Resolution 1973 (2011) Concerning Libya*, S/2014/106, February 19, 2014; Frederic Wehrey, "The Challenge of Violent Extremism in North Africa: The Case of Libya," Carnegie Endowment for International Peace, March 29, 2017; Johannes Saal, "The Islamic State's Libyan External Operations Hub: The Picture So Far," *CTC Sentinel* 10, no. 11 (December 2017).

153 Council on Foreign Relations, "Civil Conflict in Libya," *Global Conflict Tracker*,

This civil war drew in regional powers: the UAE, Egypt, and Russia backed Haftar; Turkey and Qatar supported the Tripoli government. The conflict remains unresolved, with a fragile UN-mediated ceasefire in place since 2020 but no genuine national reconciliation.

Coup d'état (Optional Branch)

Multiple attempted power grabs by rival factions and generals have punctuated Libya's post-revolutionary chaos. Haftar's 2019-2020 offensive against Tripoli can be understood as an attempted coup, a bid to unify Libya under military rule. Turkish military intervention in early 2020 prevented Haftar's forces from capturing Tripoli and collapsing the Government of National Accord, but it produced a new military stalemate rather than a durable political settlement.[154]

Summary: The Costs of Fragmentation

Lacking a shared vision for Libya's future or a compelling justification for collective action, post-revolutionary politics quickly became "hyperlocalized." In the vacuum left by the collapse of state authority, regional, tribal, and local loyalties reasserted themselves, undermining any serious attempts at forging national unity.

From the outset, Libya's revolutionary vanguard was fragmented and disjointed. Rather than being led by a disciplined, ideologically driven core, the uprising was spearheaded by a loose coalition of militias, local notables, and defected military officers, each advancing their own interests with minimal coordination.

This lack of organizational capacity and ideological cohesion made it impossible to transition from rebellion to effective governance. In the aftermath of Gaddafi's fall, power devolved to local militias, each carving out its own territory and resources, frequently clashing with rivals rather than cooperating.

Attempts by figures such as Khalifa Haftar to unify the country under a single military or political structure have consistently been stymied by the

---

updated July 15, 2024; European Council on Foreign Relations, *A Quick Guide to Libya's Main Players* (London: European Council on Foreign Relations, December 2016); Christopher M. Blanchard, "Fractured Libya: The Struggle for Unity," testimony before the House Foreign Affairs Subcommittee on the Middle East and North Africa, April 18, 2018; Human Rights Watch, *"We Feel We Are Cursed": Life under ISIS in Sirte, Libya* (New York: Human Rights Watch, 2016).

[154] International Crisis Group, *Turkey Wades into Libya's Troubled Waters*, Europe Report no. 257 (Brussels: International Crisis Group, April 30, 2020); European Council on Foreign Relations, "Libya: Into the EastMed Arena," accessed May 22, 2026.

conflicting ambitions and parochial loyalties of Libya's many factions. Alliances are often forged out of convenience or local interest, rather than any genuine commitment to a national project.

The result is a fractured political landscape where legitimacy is constantly contested, governance is deeply fragmented, and the prospects for national reconciliation remain remote. Efforts by Haftar to build a unified military structure have been hampered by the contradictory goals and parochial loyalties of component groups, with alliances often based on personal or local interests rather than a shared national project.

Ultimately, Libya's failure to develop a cohesive ideology and a unified vanguard has been central to its persistent fragmentation, allowing local, tribal, and factional interests to fill the void and preventing the emergence of a stable, unified state. More than a decade after Gaddafi's fall, Libya remains divided, impoverished, and violent, a cautionary tale of revolution without coherent vision or organized leadership.

Syria: From Peaceful Protest to Endless War

Syria's descent into civil war exemplifies the most catastrophic possible outcome of revolutionary upheaval: protracted multi-sided conflict, massive humanitarian disaster, international intervention, state collapse, and uncertain reconstruction. The RTVC framework helps explain how Syria's 2011 uprising evolved into a 13-year nightmare that only ended, or at least paused, with Assad's fall in December 2024.

Background: The Assad Regime

Hafez Assad took power in Syria through a coup in 1970. He was succeeded, in 2000, by his son Bashar, whom many hoped would bring democratic reforms. But the regime's survival rested on a coalition of minorities, Druze, Christians, Kurds, and the ruling Alawites. The coalition led by the Alawites (an offshoot of Shia Islam comprising about 12% of Syria's population) acted as a counter to the strength of the Sunni Arab majority (about 60% of the population). For the regime and its supporters, the Arab Spring represented an existential threat.

Ideological Development

As in Egypt the Muslim Brotherhood had long championed the Islamist agenda and opposed Assad's ruling Arab Socialist Ba'ath Party. But following a Brotherhood led uprising in 1982 the Muslim Brotherhood was brutally suppressed. As many as 38,000 people were killed and many of the Brothers fled the country.[155]

---

155 Yasmina Allouche, "The 1982 Hama Massacre," *Middle East Monitor*, February 12,

Beyond the Brotherhood, Syria's opposition was ideologically diverse from the outset: secular democrats, Arab nationalists, Kurdish separatists, and Islamists of various stripes. The Assad regime's Ba'athist nationalism was seen as corrupt and sectarian, serving Alawite interests rather than the Syrian nation.

However, this ideological diversity prevented the emergence of a unified revolutionary program. Different groups wanted incompatible things: federalism for Kurds, democracy for liberals, Islamic state for jihadists, return of Sunni dominance for traditionalists. This fragmentation would prove fatal to prospects for post-Assad stability.

The Vanguard

Early protests in March 2011 were led by youth and civil society activists inspired by Tunisia and Egypt. The regime's violence quickly radicalized the opposition. Militant groups emerged: first the Free Syrian Army (composed largely of military defectors), then jihadist factions like Jabhat al-Nusra (al-Qaeda's Syrian affiliate) and ISIS. Syrian Islamist returning from fight the Americans in Iraq, filled the ranks of the jihadists with experienced combatants.

These militants conducted assassinations, bombings, and provocations which elicited intense regime reaction. The vanguard's militarization and fragmentation made peaceful resolution impossible and set the stage for protracted conflict.

Mass Civil Unrest and Revolution

The uprising began with peaceful protests against Assad's regime in March 2011, inspired by the broader Arab Spring and demanding political reform, release of political prisoners, and an end to the decades-old emergency law. Nationwide demonstrations escalated into armed rebellion as the regime's brutal crackdowns radicalized the movement. Security forces fired on unarmed protesters, arrested and tortured activists (including children), and deployed the military against civilian populations.

Protesters took up arms in response, shifting the conflict from civil unrest to armed rebellion. Overt opposition coalitions formed, but clandestine cells and foreign-backed militias conducted sabotage and attacks. The regime's crackdown blurred the lines between protest and insurgency, and the conflict rapidly militarized.

By framing the uprising as a sectarian and jihadist threat rather than acknowledging legitimate grievances, Assad deepened divisions and justified further repression. The regime's strategy, create chaos, then position itself as the only alternative to that chaos, succeeded in fracturing the opposition but also destroyed Syria itself.

2018, https://www.middleeastmonitor.com/20180212-the-1982-hama-massacre/.

Revolutionary Victory and Consolidation of Power

Syria's revolution devolved into a protracted civil war lasting over 13 years. A portion of the Alawite-dominated military remained loyal, and backed by Russia, Iran, and Lebanese Hezbollah, enabled the survival of Assad's regime through 2024.

ISIS briefly established a brutal theocracy (2014-2019) controlling large swaths of eastern Syria and western Iraq, declaring a "caliphate" under Abu Bakr al-Baghdadi. However, ISIS was crushed by combined international intervention: U.S.-led coalition airstrikes, Kurdish-led Syrian Democratic Forces (SDF) ground operations, and campaigns by Assad's forces and their Russian/Iranian backers.

When Russian support diminished due to commitments in Ukrains and Both Iran and Lebanese Hezbollah suffered setback in their ongoing conflict with Syria a window opened. HTS (Hay'at Tahrir al-Sham), emerged from al-Nusra and broke with al-Qaeda in 2016. Under Ahmed al-Sharaa (formerly Abu Mohammad al-Jolani), HTS led the 2024 offensive that toppled Assad.

In December 2024, Assad's fall led to a transitional government dominated by HTS which attempted to rebrand as a more moderate force focused on governance rather than jihad. However, power struggles and sectarian tensions persist as various armed groups compete for influence. There have been massacres among the Alawite and Druze communities blamed on the new regime.[156] HTS, however, denies responsibility pointing to rogue elements which it claims to be investigating.

Tyranny

ISIS exemplified revolutionary tyranny: mass violence, cults of personality, and systematic repression. ISIS brutality included public executions, mass beheadings (including of journalists and aid workers), destruction of cultural heritage sites (Palmyra), enslavement of Yazidi women, and the establishment of a totalitarian theocracy that permitted no dissent. ISIS's propaganda videos of executions were designed to terrorize enemies and attract foreign fighters. A Coalition led by the United Stated brought an end to ISIS' caliphate but left Assad in power until the HTS offensive.

HTS and al-Sharaa are at an inflection point; will it transition to democracy or install the next tyranny? Early signs are mixed: the transitional government has promised inclusive governance, protection for Syria's diverse religious minorities including Christians, Druze, and Alawites, and a gradual transition to civilian rule. However, concerns persist about HTS's Islamist ideology, its capacity for genuine democratic governance, and whether it will pursue

---

[156] Güneş Murat Tezcür, "Understanding the Violence against Alawites and Druze in Syria after Assad," *The Conversation*, July 24, 2025,

reconciliation or retribution.

Outbreaks of violence, score-settling against former regime officials, and the question of how to handle the thousands imprisoned by Assad's security apparatus continue to test Syria's fragile post-Assad order. The international community's response, particularly regarding sanctions relief and reconstruction aid, will significantly influence whether Syria can escape the cycle of tyranny or whether new forms of authoritarian rule will emerge under al-Sharaa's hand.

Export of Terror and Expansion

Syria became the epicenter for transnational jihadism, exporting fighters and ideology across the region and beyond. The conflict drew in foreign powers and non-state actors, amplifying its regional impact.

Foreign fighters from over 80 countries joined the Syrian conflict, with tens of thousands flocking to join ISIS, al-Nusra, and other groups. Returnees from Syria have conducted or inspired terrorist attacks in Europe, the, and across the Muslim world.

The Syrian conflict also became a training ground for the next generation of jihadists, much as Afghanistan in the 1980s spawned al-Qaeda. Networks established in Syria continue to operate globally.

Off-Ramp to Democratic Transition

Attempts at negotiated transition repeatedly failed due to external intervention, sectarian fragmentation, and the dominance of armed actors. Multiple rounds of Geneva talks, Astana process negotiations, and UN-mediated efforts produced ceasefires that quickly collapsed but no political settlement.

The off-ramp remains elusive, though the current transitional government claims to aspire to legitimate governance. Whether Syria can achieve a genuine democratic transition or will cycle back to authoritarianism depends on factors including: the transitional government's willingness to share power, the behavior of armed groups, the response of Syria's minorities (especially Alawites fearful of retribution), and the role of external powers (Turkey, Iran, Russia, the United States).

Civil War (Optional Branch)

Syria's revolution became a textbook case of civil war, with multiple factions controlling territory, foreign intervention, and deep societal splits.

The conflict involved: Assad's forces (backed by Russia, Iran, Hezbollah); ISIS (until 2019); Kurdish forces/SDF (backed by the U.S.); various Free Syrian Army factions; al-Nusra/HTS; Turkish-backed opposition groups; and numerous smaller militias. At its peak, the war had dozens of active fronts with shifting alliances.

The war claimed an estimated 500,000-600,000 lives (direct combat deaths plus indirect deaths from disease, lack of medical care, and starvation), displaced over 13 million people (more than half of Syria's pre-war population of 22 million), created over 6 million refugees who fled to Turkey, Lebanon, Jordan, and Europe, and destroyed much of Syria's infrastructure, housing, and cultural heritage.

Coup d'état (Optional Branch)

Assad's regime survived multiple coup attempts over the decades, but the final collapse in December 2024 was precipitated not by a traditional military coup but by a swift insurgent offensive combined with catastrophic loss of external support.

The Timeline of Collapse:

- September-October 2024: Israeli airstrikes systematically eliminate Hezbollah's senior leadership, including Secretary-General Hassan Nasrallah, decimating the organization's capacity to support Assad
- November 2024: Russia, deeply committed to its war in Ukraine, reduces military support to minimal levels; Iranian proxies across the region have been weakened by sustained Israeli operations
- November 27, 2024: HTS launches offensive from Idlib province; regime forces, demoralized and unpaid, offer minimal resistance
- November 29-30: Aleppo, Syria's largest city, falls to rebel forces with stunning speed
- December 1-7: Rebel advances accelerate as regime forces dissolve; major cities fall in rapid succession: Hama (December 5), Homs (December 7)
- December 8, 2024: Rebels enter Damascus without major battle; Assad flees to Russia; regime collapses after 13 years of civil war

The rapidity of the collapse shocked observers worldwide. What had appeared to be a frozen conflict shifted decisively once Assad's foreign backers could no longer sustain him. The regime's brittleness, masked for years by external support, became catastrophically apparent.

Summary

Syria's descent into civil war was fueled by the interplay of revolutionary violence and counterrevolutionary repression, creating a spiral that lasted over 13 years. The regime's indiscriminate violence and sectarian tactics aimed to crush dissent, but instead hardened opposition resolve and encouraged militarization.

These dynamics, regime violence legitimizing armed resistance and the blurring of lines between protest and insurgency, fractured Syrian society and made peaceful resolution impossible. The result was a spiral into full-scale, multi-sided civil war that drew in regional and international powers, generated millions of refugees, and claimed hundreds of thousands of lives.

Assad's regime survived until December 2024, but its eventual collapse demonstrated that coercion alone cannot indefinitely suppress the demand for justice and dignity. The transitional period now underway will test whether Syria can escape the revolutionary cycle and achieve the democratic aspirations that sparked the 2011 uprising, or whether new forms of tyranny will emerge from the ashes of the old regime. The original grievances, sectarian inequality, economic hardship, political repression, remain largely unaddressed, and Syria's future hangs in the balance between fragile hope and the specter of renewed conflict.

COMPARATIVE ANALYSIS

Having examined each country's revolutionary trajectory through the RTVC framework, we now turn to cross-cutting themes that help explain divergent outcomes.

The Role of Military and Security Forces

The reaction of the military proved important to revolutionary outcomes may determine whether there is civil war or not. Albrechts *Military Insubordination in Popular Mass Uprisings*, stresses the importance of studying not loyalty but rather insubordination in the military, and in particular how popular uprisings affect military insubordination.[157]

Albrecht classifies types of insubordination by agency (Commander or Combatant), by collective action (either Coup or Mutiny), and by individual action (either Defection or Desertion). He defines a coup as "illegal and overt attempts by the military…to unseat the sitting executive."

The behavior of military and security forces was decisive in shaping revolutionary outcomes:

Tunisia: The military refused to fire on protesters and remained neutral throughout the uprising, enabling a peaceful transition. The Tunisian military's small size, professional ethos, and lack of economic interests in maintaining the Ben Ali regime contributed to its neutrality. This neutrality was critical to preventing both state collapse (as in Libya) and military coup (as in Egypt).

Libya: Military fragmentation and defection led to state collapse and militia rule. Parts of the military defected to the rebellion, other parts remained loyal to Gaddafi, but no unified military institution survived the conflict. The result was not military rule but military dissolution, leaving militias as the dominant armed actors.

---

[157] Albrecht, "Military Insubordination."

Egypt: The military acted as kingmaker, first facilitating revolution by refusing to massacre protesters in Tahrir Square, then crushing the Muslim Brotherhood to restore its own power. The Egyptian military's extensive economic interests—variously estimated by analysts at roughly 15 to 40 percent of the economy, though precise figures remain contested—its corporate identity, and its historical self-conception as guardian of the state motivated it to preserve the system while sacrificing individual leaders when necessary.[158] SCAF managed the transition to protect military prerogatives, then intervened via coup when the Brotherhood threatened those interests.

Syria: The military's loyalty to Assad, bolstered by sectarian alliances (Alawite officers commanding largely Sunni conscripts), enabled the regime's survival and brutal repression for 13 years. Defections occurred but were insufficient to fracture the core command structure. Russia and Iran's support reinforced this loyalty with material backing. Only when that external support collapsed in 2024 did the military's cohesion dissolve, leading to rapid regime collapse.

Albrecht's analysis of military insubordination highlights the importance of whether commanders or rank-and-file defect, mutiny, or remain loyal, shaping whether revolutions lead to civil war, coup, or peaceful transition.

Synthesis: Military behavior is perhaps the single most important variable in determining revolutionary outcomes. Where the military stayed neutral (Tunisia), democracy had a chance. Where it fragmented (Libya), state collapse ensued. Where it remained cohesive and self-interested (Egypt), it imposed its own solution via coup. Where it maintained loyalty through sectarian bonds and external support (Syria), protracted civil war resulted.

Islamist Movements: Strategies and Outcomes

The Arab Spring tested competing Islamist strategies: gradual reform versus revolutionary jihad, democratic participation versus armed struggle, ideological purity versus pragmatic compromise.

Tunisia - Ennahda's Moderation: The Ennahda party demonstrated willingness to compromise, share power, and operate within democratic norms. It won elections but agreed to coalition governments, stepped down when its policies proved unpopular, and accepted secular provisions in the Constitution. This moderation was crucial to Tunisia's democratic survival, though Ennahda

---

158 Kenneth M. Pollack, "The Egyptian Military Faces Its Defining Hour," Brookings Institution, November 28, 2011; Yezid Sayigh, *Above the State: The Officers' Republic in Egypt* (Washington, DC: Carnegie Middle East Center, 2012); Silvia Colombo, "The Military, Egyptian Bag-snatchers," *IAI Working Papers* 14, no. 5 (November 2014); William Parsons and William Taylor, *Arbiters of Social Unrest: Military Responses to the Arab Spring* (Colorado Springs, CO: United States Air Force Academy, 2011).

has since faced criticism for being too accommodating thus losing political support.

Egypt - The Brotherhood's Overreach: The Muslim Brotherhood's attempt to monopolize power after winning elections alienated coalition partners, frightened minorities and secularists, and provided justification for military intervention. The Brotherhood's history of persecution may have created a siege mentality, leading it to grab power while it could rather than consolidate gradually. This strategic error proved fatal.

Libya - Jihadist Opportunism: The collapse of state authority allowed various jihadist groups (ISIS, Ansar al-Sharia) to establish territorial control temporarily. However, their extremism and brutality ensured they could not build lasting legitimacy, and they were eventually pushed out by rival militias and international intervention.

Syria - Radicalization and Fragmentation: Syria saw the full spectrum of Islamist strategies. Moderate groups initially advocated political reform. As violence escalated, radical groups (al-Nusra, ISIS) gained prominence through superior military effectiveness and ideological clarity. HTS's eventual 2024 victory represents a potential middle path: former jihadists attempting to rebrand as pragmatic Islamists. Whether this happens remains uncertain.

Synthesis: Islamist movements succeeded where they demonstrated flexibility, moderation, and willingness to share power (Tunisia). They failed catastrophically where they pursued ideological purity and monopolization of power (Egypt, ISIS in Syria/Iraq). The question now is whether HTS in Syria can replicate Ennahda's successful moderation or will revert to exclusionary Islamism.

External Intervention: Helping or Hindering?

External intervention played complex and often contradictory roles across the Arab Spring cases.

Libya - Intervention's Pyrrhic Victory: NATO's 2011 intervention prevented a potential massacre in Benghazi but contributed to Libya's long-term fragmentation. By enabling dozens of militias to succeed against Gaddafi without coordinating under unified command, NATO inadvertently ensured post-war chaos. Western powers declared victory and departed, leaving Libyans to cope with state collapse and inadvertently triggering a migrant crisis which continues to tax European humanitarian values.

Syria Intervention's Failures and Contradictions: Multiple interventions by various actors produced devastating results:

- U.S.-led coalition against ISIS succeeded militarily but did little to address underlying conflict
- Russian intervention (2015) decisively supported Assad but at

tremendous humanitarian cost

- Iranian/Hezbollah support sustained the regime through its darkest hours
- Turkish intervention backed various opposition groups but prioritized containing Kurdish autonomy over broader political settlement
- The net result: foreign powers pursuing contradictory agendas perpetuated stalemate and suffering

Egypt - Ambivalent Support: Western powers, particularly the U.S., sent mixed messages: supporting democracy rhetorically while maintaining close ties to the military. When Sisi seized power, Western response was muted, prioritizing stability and counterterrorism cooperation over democratic values. This tacit acceptance of the coup demonstrated the limits of democracy promotion rhetoric.

Tunisia - Benign Neglect: Relatively limited external intervention in Tunisia may have been a blessing. Tunisia received some economic support and diplomatic encouragement but was largely left to work out its own path. This allowed organic, locally driven solutions to emerge.

External intervention in the Arab Spring generally exacerbated problems rather than resolved them. Well-intentioned intervention (Libya) produced unintended consequences. Self-interested intervention (Syria) perpetuated conflict. Rhetorical support for democracy (Egypt) proved hollow when confronted with realpolitik. Only Tunisia, which received the least directive intervention, achieved partial success. The lesson: sustainable political systems must emerge from indigenous social forces, not external imposition.

AFTERMATH AND ONGOING STRUGGLES

The "Arab Winter"

The early optimism of the Arab Spring quickly gave way to what many analysts call the "Arab Winter," a period marked by violence, authoritarian resurgence, and deepening sectarian conflict, especially between Sunni and Shia Muslims. Weakened state institutions and power vacuums allowed extremist groups like ISIS to flourish, further destabilizing the region and exporting violence beyond its borders.

Syria's civil war (2011-2024) became the most devastating manifestation of this period, claiming an estimated 500,000-600,000 lives, displacing over 13 million people (more than half of Syria's pre-war population), and destroying much of the country's infrastructure and cultural heritage. The conflict drew in regional powers (Turkey, Iran, Saudi Arabia) and global actors (Russia, The United States), becoming a proxy war that internationalized what began as a domestic uprising. Chemical weapons use, mass atrocities, and systematic torture characterized the conflict's brutality.

Even with Assad's December 2024 defeat, Syria's future remains deeply uncertain. The transitional government grapples with monumental challenges: reconstructing a devastated country, reconciling deeply divided communities, integrating former combatants from dozens of armed groups, determining accountability for war crimes, managing the return of millions of refugees, and preventing renewed conflict. The risk of state fragmentation renewed civil war, or the emergence of a new authoritarian regime remains substantial. Syria exemplifies how revolutionary upheaval can lead not to democratic transition but to protracted state collapse and humanitarian catastrophe.

Revolutionary Logic: Where Does It Go From Here?

Groups like Al Qaeda and the Islamic State, drawing on Che Guevara's *foco* theory, sought to act as revolutionary vanguards, using dramatic violence to inspire mass mobilization. For a time, ISIS succeeded in seizing territory and rallying followers. However, with its territorial defeat in Syria and Iraq (2017-2019), the radical Islamist project has reverted to earlier stages: clandestine organization, small-group action, and ideological groundwork.

The Islamic State expanded its influence into Libya, Nigeria, and Yemen during its peak years (2014-2017), establishing territorial control in parts of Libya and creating affiliates across multiple countries under the ISIS brand. It exported terror to Europe and the United States, claiming responsibility for attacks in Paris (November 2015, January 2015), Brussels (March 2016), San Bernardino (December 2015), and Orlando (June 2016). Through sophisticated social media operations, ISIS exported its ideology worldwide, recruiting foreign fighters from dozens of countries and inspiring "lone wolf" attacks far from its physical territory.

Following ISIS's territorial defeat, the organization reverted to insurgent tactics. While no longer controlling significant territory, ISIS maintains a reduced but persistent presence through affiliates and cells, particularly in:

- Syria and Iraq: Continued low-level insurgency in rural areas
- West Africa: ISIS-West Africa Province (ISWAP) remains active in Nigeria and the Lake Chad region
- Afghanistan: ISIS-Khorasan Province (ISIS-K) conducts attacks against the Taliban government
- Other regions: Smaller affiliates in Mozambique, Democratic Republic of Congo, and parts of Southeast Asia

Post-2024, with Assad's fall, concerns have emerged about potential ISIS resurgence in Syria's post conflict turmoil. The transitional government's control remains incomplete, and thousands of ISIS fighters remain detained in camps controlled by Kurdish forces, representing a continuing security challenge.

Some Islamists are now turning their energies toward "educating" and mobilizing the masses, shifting the battlefield from open conflict to the realm of hearts and minds. In Egypt, the Muslim Brotherhood's brief experiment with power ended in failure, driving some members toward non-violent movements like *Dawah and Tabligh*. These groups, unlike their militant counterparts, focus on spiritual renewal and moral reform, arguing that societal injustice stems from a loss of Islamic values. As one da'wah activist in Syria put it, "Jihad is not waged on the battlefields; it is waged in the hearts of men and women, Muslims and non-Muslims alike. Dawah is jihad."[159] Such movements, rejecting Western secularism and political activism, create enclaves that are difficult for states to counter and are gaining influence in countries like Turkey and Qatar.

Regional Power Struggles

The idea of an Islamic state, as re-envisioned by Muhammad Rashid Rida after the Ottoman collapse, was less about reviving the caliphate and more about forging a new political order—perhaps a Turko-Arab partnership with a spiritual leader at its core. Today, Turkey's assertive role in Syria and the broader region can be seen as a bid for Islamic leadership, offering a potential alternative to both jihadist chaos and Western-imposed models. However, Turkish ambitions are met with suspicion by both Western and Arab states and are actively opposed by powers like Russia and Iran.

Iran, leveraging its "Axis of Resistance" (including Hezbollah, Iraqi militias, and formerly Assad), positioned itself as the defender of Shia Muslims and opponent of Western hegemony. However, Israeli operations throughout 2023-2024 severely degraded this network, culminating in the elimination of Hezbollah's leadership and the collapse of its Syrian ally. And, as of this writing the Islamic Republic finds itself isolated as it faces the wrath of American power.

Saudi Arabia and the UAE led a counter-revolutionary push, supporting military rule in Egypt and various factions in Libya and Yemen to prevent Islamist gains. Their vision emphasizes authoritarian stability over democratic experiments.

Social and Economic Impact

The Arab Spring was fueled by demands for dignity, human rights, and economic opportunity. Yet the aftermath in many countries has brought increased repression, economic hardship, and mass displacement. Millions have been forced to flee their homes, particularly from Syria, Libya, and Yemen, creating one of the largest refugee crises in modern history. The resulting

[159] Taylor Luck, "The Mysterious Islamic Movement Quietly Sweeping the Middle East," *Christian Science Monitor*, December 6, 2015.

migration has strained neighboring countries (Turkey, Lebanon, Jordan) and sparked political debates across Europe over refugee acceptance and integration.

Economic conditions have worsened in most Arab Spring countries. Youth unemployment remains crushingly high. Corruption persists. The potential for further unrest has not yet abated.

Legacy and Continuing Unrest

While the Arab Spring failed to deliver widespread democracy, it fundamentally altered the region's political landscape. It exposed the fragility of authoritarian regimes and demonstrated the power of popular mobilization, even if that power was subsequently suppressed. The Arab Spring showed that long-entrenched dictators could be toppled and that ordinary citizens could challenge state power.

Yet the core grievances that fueled the uprisings remain largely unresolved. This has led to new waves of protest, sometimes referred to as a "Second Arab Spring":

- Algeria (2019): Mass protests forced President Abdelaziz Bouteflika to resign after 20 years in power
- Sudan (2019): Protests toppled Omar al-Bashir after 30 years of rule
- Iraq (2019-2020): Protests against corruption and Iranian influence
- Lebanon (2019-2020): Protests against economic collapse and sectarian political system

These new movements learned from the first Arab Spring's failures, emphasizing sustained pressure, clear demands, and awareness of co-optation risks. Yet they also face familiar obstacles: entrenched elites, military intervention, external interference, and economic crisis.

Public opinion across the region reflects disappointment with the outcomes but also persistent desire for reform and accountability. Surveys consistently show majorities supporting democracy in principle while expressing skepticism about its feasibility given current conditions.

CONCLUSION

For many in the region, the Arab Spring is not a closed chapter but an ongoing process. Syria's December 2024 transition from Assad's brutal rule to an uncertain future exemplifies this continuing struggle. The revolutionary demands for justice, dignity, and self-determination that erupted in 2011 persist, even as the forms of struggle and governance evolve. Whether Syria's transition will lead to the democratic aspirations that sparked the Arab Spring, or to new iterations of authoritarianism, whether Islamist or military, remains to be determined.

The revolutionary wave is neither a single, unified movement nor a simple failure. Instead, it is a decades-long struggle, marked by cycles of hope, repression, and renewed activism. Tunisia's fragile democracy, Egypt's return to military rule, Libya's ongoing fragmentation, and Syria's uncertain transition together illustrate the diverse trajectories possible when revolutionary fervor meets entrenched power structures.

The Ongoing Quest for Justice and Dignity

As the Middle East continues to grapple with shifting geopolitical dynamics and persistent unrest, the ultimate outcome remains uncertain. The West must learn to respect the agency of the region's peoples, accepting that their choices, whether toward democracy, Islamism, or hybrid systems that blend tradition and modernity, are theirs alone to make.

The true legacy of the Arab Spring may yet be written in:

- Syria's reconstruction and whether it can achieve inclusive governance
- Egypt's reckoning with military rule
- Libya's potential stabilization if factions find common ground or a permanent splint into East and West Lybia
- Tunisia's defense of its democratic gains against authoritarian backsliding
- The ongoing protests from Algeria to Iraq to Lebanon

The struggle for justice, dignity, and self-determination in the Arab world has not ended; it has merely entered a new phase, shaped by the hard lessons of the past thirteen years. The RTVC framework shows us that revolutions follow predictable patterns, but also that off-ramps exist, if actors are willing to compromise, if civil society is strong, if militaries show restraint, and if external powers allow indigenous solutions to emerge.

The Arab Spring demonstrated both the possibility and the peril of revolutionary change. Its ultimate legacy remains unwritten, dependent on choices yet to be made by the region's people, leaders, and external actors. What is certain is that the demand for justice and jihad will continue to shape the Middle East's political landscape for generations to come.

CHAPTER 3

REVOLUTIONS AND CONSTITUTIONS: The English, American, And French Revolutions Compared

The revolution of the United States was the result of a mature and reflecting preference for freedom, not of a vague or ill-defined craving for independence. It contracted no alliance with the turbulent passions of anarchy, but its course was marked, on the contrary, by a love of order and law.
–Alexis de Tocqueville, *Democracy in America.*

The Republican Revolutions of the 17th and 18th Centuries represent a fundamental restructuring of governmental authority and, more to the point, a new conceptualization of the relationship between a government and the governed. Perhaps it is not correct to say 'new conceptualization" the word republic comes from the Latin "res publica" or "public affairs" as used by the Romans.[160] However, from the time the Caesars successfully asserted power over the Roman Senate until the Reformation, kings and emperors ruled most of Western Civilization. The Republican Revolutions reversed that situation.

The Revolutions in America and France during the 18th Century were, in part, products of Enlightenment thinking. The Enlightenment, a period of philosophical thought, was one of the great turning points in Western Civilization marking a definitive end to absolute monarchy and the rise of constitutional democracies. The Enlightenment, however, could not have occurred without the preceding Reformation.

Enlightenment ideals justifying revolution can, arguably, be traced to Luther's revolt against the Roman Catholic Papacy. The Reformers did not see themselves so much as revolutionaries as much as they did reactionaries. Before

---

[160] Jack A. Goldstone, *Revolutions A Very Short Introduction*, (New York: Oxford University Press, 2014), 44.

Emperor Constantine (280-337A.D.), the church had been generally decentralized albeit with a number of leading bishops in important cities such as Antioch, Alexandria and of course Rome. Constantine's conversion and subsequent influence over the church led to its' institutionalization and the elevation of the Roman papacy to a position of real power. In the 11th Century the church divided between the Orthodox Church centered in Constantinople and The Catholic Church centered in Rome.

In 1453 Constantinople fell into the hands of the Ottoman Empire marking the end of the Byzantine, or East Roman, Empire making relations between the Orthodox and Roman Churches even more difficult. To survive the chaos following the fall of the West Roman Empire in the 5th Century, the Roman church had become involved in politics. Indeed, Western Europe referred to itself as Christendom distinguishing itself from the lands under Muslim domination. By the 1500s, the Hapsburg dynasty ruled a large portion of Western Europe. In return for the legitimating title of Holy Roman Emperor, the Hapsburgs provided security for the Roman Bishop who had become known as the Pope. The Popes, intimately involved in secular and spiritual matters, were sometimes pious and other times less so. In addition to crowning the Hapsburg Emperors, the Roman Popes, claiming divine authority, legitimized monarchs throughout Europe.

This fusion of spiritual and secular authority had never gone unchallenged. Jan Hus had died at the Council of Constance in 1415 for demanding precisely the return to first principles that Luther would demand a century later, and Hus was neither the first nor the last to make the attempt. What the early sixteenth century provided that earlier reformers had lacked was a confluence of pressures, a printing press capable of spreading dissent faster than councils could suppress it, a rising literate class with grievances against Rome's financial demands, and princes with political incentives to shelter reformers the Pope would otherwise have burned.

Luther and his followers called for reform of the church and a return to first principles, or the fundamentals of the faith. They transformed society and the relationship between individuals and the institutional church and between the Government and the governed. Protestant reformers emphasized man's connection to the Almighty. Man is first created in God's image, and therefore of exceedingly more value than other creatures. Second, because he is God's special creation, man has direct access to the creator with little need for the clergy's intermediary role. Protestant theology recognized man's inherent sinfulness and the propensity to abuse positions of power for gain. This led to the conviction that the Pope had no greater hold on the word of God than any other individual.

Those supporting the Pope's rule argued that there must be a central standard maintained by an established authority. Failure to do so would result in chaos as a Pandora's Box filled with unchecked heresy opened. The Protestants recognized this danger but argued having a single point of failure, the Pope, meant that if he was corrupt, then the entire church was corrupted. Allowing each to choose his own way meant while some would be corrupt, others would not, and God would recognize and guide his sheep.

This democratization of Western Christendom brought into question ideals of divine right and absolute Monarchy. Similar arguments were made for and against Monarchy: In the same way, monarchists argued a central and established system or ruling class of elites was best to maintain order, as this class was trained and uniquely formed to govern. Allowing the masses too much autonomy was likely to result in anarchy.

Reformation ideas, coupled with the power of Gutenberg's press, opened the way for the emergence of the Enlightenment in the earlier modern period. With growing access to reading material and expanding literacy, Ideas spread rapidly. Opportunity blossomed as well. Knowledge being power, the Elites became less elite as knowledge became widely available.

While many Enlightenment writers championed liberal secularism, such ideas could not have found freedom of expression before the Reformation. It was during the wars of the Reformation that Western thinkers began theorizing as to whether resistance and revolution against their governments were justified.

The justification for resistance to unjust rulers relied heavily upon the rights of individuals. Those rights belong to individuals because they are each created in the image of God, inalienable rights endowed by their creator. Yet, importantly, Enlightenment writers also pursued the balance between order and oppression, posing the question, what is the point at which Government becomes so oppressive that the need to revolt and the resulting violence overrides the desirability of stability, peace, and order?

England answered that question twice in the same century, first through civil war and regicide, then through the constitutional settlement that became the model for every Anglo-American republic that followed. In both cases the trigger was not abstract ideology but the accumulated weight of economic hardship, religious conflict, and a crown that had pushed the balance between order and oppression past the point the governed were willing to bear.

ENGLISH CIVIL WAR AND THE GLORIOUS REVOLUTION

The poorest {man} that is in England hath a life to live as the greatest {man}.
–Thomas Rainsborough, The Putney Debates, 1648[161]

---

[161] John Rees,
The Leveller Revolution: Radical Political Organization in England, 1640–

REVOLUTIONS and CONSTITUTIONS

England in the 17th Century was undergoing significant change. The population had more than doubled from the proceeding Century, and agriculture, while reforming, was not keeping pace with those it had to feed. Additionally, a significant expansion in trade made exporting grain a lucrative proposition compounding the tightening resources leading to numerous food riots. The expanded population resulted in other challenges as well. Land ownership became increasingly restricted due largely to the division of inherited land; a man would divide his land between his sons, who in turn would divide their land between their sons so that family parcels became increasingly smaller. Alternatively, a man would transfer all his land to his eldest son leaving the others destitute.

Few options for personal advancement remained available to such men. Service to the King or Church became highly sought after with more applicants for the limited positions available. A large portion of the population, having no property in the rural areas, moved to the cities. The increase in population meant cheaper labor and decreased wages. Unemployment, food scarcity, and shrinking wages contributed to fewer opportunities to better one's lot in life. Simultaneously, the Crown was largely bankrupt due to war, and the nobility were defending their autonomy from the King.

The dispute between English Kings and Parliament goes back to at least 1215 when English Nobles rebelled against King John's tyrannical and arbitrary rule. The King was forced to sign the Magna Carta, which, among other things, called for due process under the law, guaranteed certain property rights for the English Barons, and the consent of the nobility to taxation.[162] As a result, English Kings generally did not interfere in local rule other than collecting taxes, mostly in times of war. The Magna Charta allowed for the formation of Parliament. When the King needed to raise funds, he had to call on the nobles to consent; this would become the House of Lords.

In 1258, during the reign of Henry III, the provisions of Oxford attempted to make the King accountable to the Barons.[163] The nobles and the King chose the King's ministers jointly through Parliament. The King retained the right to call and dissolve Parliament. In 1265, one of the Barons, Simon de Montfort, reformed the Parliament further by forming the first Parliament with representatives from each town in England; this became the House of

---

1650 (London: Verso, 2016), 1; Thomas Rainsborough, speech at the Putney Debates, 29 October 1647, as recorded in *The Putney Debates of 1647: The Army, the Levellers and the English State*, ed. Michael Mendelson (Cambridge: Cambridge University Press, 2001), 53.

[162] The National Archives UK, "Magna Carta, 1215," last modified March 3, 2025. See clauses 12 and 14 regarding taxation and consent.

[163] Provisions of Oxford (1258), *Encyclopaedia Britannica*, last updated July 19, 1998.

Commons.[164]

In 1517 Martin Luther nailed his 95 theses to the Church doors and started the Protestant Reformation. Guttenberg's 1439 printing press with moveable type had provided a technological media revolution. Combining the ideas of the Reformation with the ability to mass-produce literature, the Protestants spread not only Bibles in native languages but also encouraged literacy for the masses so that they might read the translated scriptures.

King Henry VIII and his wife, Catherine of Aragon, were loyal Catholics until the Queen failed to produce a male heir. Refused a divorce from the Pope, Henry, in effect nationalized the Catholic Church in England. The Church of England made Protestantism acceptable and created an ongoing struggle to return a Catholic monarch to the throne and bring England under to Roman authority.

Ideological Development

The English Civil War was a twofold struggle; for secular power and over religious toleration. Within each aspect of the struggle, there were at least three subdivisions. In the earthly realm, there were struggles between the King and Parliament, between the House of Lords and the House of Commons, and between monarchists and republicans. In the spiritual realm, there were struggles between the Catholics and Protestants, between the Church of England and other Protestants, and between Presbyterians and Independents.[165]

Following the short reign of his son Edward, Henry's daughter by his first marriage, Mary, a devote Catholic, ascended to the Crown. She earned the moniker "Bloody Mary" for her unyielding suppression of Protestants, burning some 300 at the Stake.[166] English Protestants fled to the Netherlands, Germany, and Switzerland. The Scotsman John Knox fled to Geneva, where he consulted John Calvin who led the church in Geneva. Knox wrote *A Faithful Admonition to the Professors of God's Truth in England.*[167] In it he declared Queen Mary a traitor to England, likening her to Jezebel, the bloody idolatrous Old Testament queen of Israel, and calling for her overthrow.

---

[164] Sophie Ambler, "Simon de Montfort's 1265 Parliament and Magna Carta," *History of Parliament Blog,* January 19, 2015.

[165] The Independents were denominations such as Baptists and individual congregations, sometimes called sectarians or gathered churches, which were not under the authority of the national church. The Anglicans being the Church of England and the Presbyterians the Church of Scotland (generally speaking).

[166] "Mary I Queen of England," Britannica, Mary I | Biography & Facts | Britannica.

[167] "John Knox: A Faithful Admonition to the Professors of God's Truth in England," *The Tudor Society,* July 20, 2015.

It was, arguably, Catholics that were most noted in their attempts to overthrow the Kings and Queens of England. Mary's Protestant sister Elizabeth succeeded her. The Pope himself called for her assassination, and indeed several attempts were made.[168] Elizabeth never married and had no successors; she was the last of the Tudor line. James, the Protestant son of Elizabeth's Catholic rival Mary Queen of Scots, received the Crown of England and united the Kingdom with Scotland and Ireland. James I was a proponent of absolute rule as was his son and successor, Charles I.

Henry Martin was a member of the House of Commons and a proponent of republican ideals. An anti-royalist, Marten, stated, "I do not think one man wise enough to rule us all."[169] An early advocate for "people power," he felt Parliament represented and ought to receive instructions from the people, not the King. He acted on his beliefs and, during the civil war, "liberated" property from royalists and nobles for the people. These "Robin Hood" acts were forceful and often violent.[170]

One of the most influential radicals of the period was "Freeborn John" Lilburne. He was known for political theater, using prison and trials before the court as platforms to espouse his ideas.[171] He presented personal injustice as indicative of broader injustices in society.[172] As an independent, he called for freedom of religion; as a commoner, he called for equality under the law.

Lilburne became the leader of a movement known as the Levellers. The name was meant as an insult; their detractors claimed they wanted to level society.[173] The Levellers wanted to level the playing field. All English men, regardless of political or social standing, should be treated equally under the law.

Lilburne and the Levellers called for a constitution setting limits on

---

[168] "Brutum fulmen, or, The bull of Pope Pius V concerning the damnation, excommunication, and deposition of Q. Elizabeth as also the absolution of her subjects from their oath of allegiance, with a peremptory injunction, upon pain of an anathema, never to obey any of her laws or commands : with some observations and animadversions upon it / by Thomas Lord Bishop of Lincoln ; whereunto is annexed the bull of Pope Paul the Third, containing the damnation, excommunication, &c. of King Henry the Eighth." In the digital collection Early English Books Online Collections. https://name.umdl.umich.edu/A69677.0001.001. University of Michigan Library Digital Collections. Accessed October 18, 2025.

[169] John Rees, *The Leveller Revolution: Radical Political Organisation in England, 1640-1650* (London: Verso, 2016), 88-91.

[170] Ibid, 238-244.

[171] Ibid., 41.

[172] Ibid.

[173] Ibid.

governmental authority and protecting individuals' right to religious and political freedom.[174] The Levellers also expressed the right of the people to take up arms against tyranny.

The Vanguard

Politically motivated clandestine terror cells did not exist during this time in history. There were, however, plots carried out by small groups. The 1605 Gunpowder plot, an attempt to assassinate the Protestant King James I by Guy Fawkes and his co-conspirators, provides an example of the many intrigues for control of the Crown with first Protestants then Catholics forced underground.[175]

Under Charles I, Law required Englishmen to be members of the Church of England.[176] Those who did not attend weekly services were reported and sometimes fined or imprisoned. Dissenters became known as independents or Sectaries. These dissenters attended unsanctioned clandestine services, even going about in disguise to avoid the authorities.[177] They demanded the right to choose their own pastors rather than those appointed by the state-controlled Bishops. They also used the unauthorized "Geneva" Bible rather than the authorized King James Version.[178]

There had, throughout the reformation period, been the smuggling of Bibles and other printed reformation material from Protestant strongholds in Holland and Geneva. Clandestine printing and distribution operations also developed on English territory. These operations became increasingly politically oriented as the independents and sects used this form of mass media to disseminate their religious beliefs and to call for religious tolerance under the law. The calls for tolerance developed into calls for judicial reforms as dissenters were thrown into prison without due process. The Levellers operated perhaps highly organized networks in order to disseminate their ideas and expand their support base. These networks included underground printing presses operated from safe houses and clandestine distributors selling the printed manifestos.

Mass Civil Unrest As A Political Tool: The Use Of Mobs And Revolution

The Duke of Buckingham was the closest advisor first to Charles I. Buckingham was incompetent and widely disliked. He had led a number of

---

[174] Rees, *The Leveller Revolution,* 15-40.

[175] "Guy Fawkes and the Gunpowder Plot," *Historic Royal Palaces*, Tower of London, August 5, 2025.; *The Gunpowder Plot and Lord Mounteagle's Letter*, Project Gutenberg.

[176] The National Archives UK, "Magna Carta and the Act of Uniformity," March 3, 2025.

[177] Rees, 60-70.

[178] University of Glasgow, "Embroidered Genevan Bible," December 31, 2005.

military campaigns with disastrous results. In addition to the military damage done to England, his influence on foreign policy led Parliament to demand the King divest himself of Buckingham. Charles would not accept what he saw as Parliament bossing him around and charged Buckingham with yet another military expedition in 1628. Before Buckingham could leave, a disgruntled army officer assassinated him. The officer, John Felton, stated he had committed the act because Buckingham had harmed the country by offering poor advice to the King. Through three months of interrogations to uncover a suspected conspiracy, Felton maintained he had acted alone.

Following Buckingham's death, relations between the King and Parliament eroded further. To avoid interference by nobles, the King had not called on Parliament in 11 years. Charles imposed the *Book of Common Prayer* on the Presbyterian dominated Scottish church causing a rebellion. When the Scots invaded, Charles needed Parliament to help levy taxes and raise an army. The nobles took the opportunity to assert their power over the throne.

Two of the King's advisors came to represent many of the Political and Religious points of contention. The representative of religious problems was Archbishop Laud, the head of the Church of England. Laud zealously suppressed religious dissenters, promoting the highly structured Anglican confession. Many Puritans thought Laud too closely aligned with Catholic spiritual practices. It was the promotion of the Anglican Common Book of Prayer, which prompted the Presbyterian dominated Scots to revolt and invade England in 1639. Many of the parliamentarians were Calvinist Puritans ever suspicious of plots to restore Catholicism to England. The King lost the dispute with the Parliament, and the Archbishop lost his head.

Thomas Wentworth, the Earl of Strafford, a leading member of the House of Commons, represents the Political problems. The King had sent him to Ireland to raise an army that could fight the invading Scottish Army. Strafford's support for the King gained him enemies in Parliament, who accused him of advising the King to use the Irish army to subdue Parliament. Backed up by riotous mobs, Parliament maneuvered the King into ordering Strafford's execution for treason.

While the nobles continued to be the primary political leaders, political participation expanded with the emerging merchant class. At times, leaders used the mobs, such as those involved in food riots, to pressure the King. What had once been spontaneous unrest began to be used as a political tool. In retaliation for Parliament's obstruction, Charles attempted the arrest of five members of the House of Commons. While the members escaped, Charles action precipitated the Civil War in 1642.

Revolutionary Victory And Consolidation Of Power

The Monarchist Army, known as Cavaliers, was generally made up of peasants led by nobles. The Parliamentarian Army was generally made up of town militias and was mostly middle-class and Puritan. The Parliamentarians were known as Roundheads because of their cropped hair style; the Royalists tended to dress flamboyantly and wear their hair long. Oliver Cromwell quickly became the leading General of the Roundheads. The Parliamentary Army was initially raised through the mass levy of enthusiastic but poorly trained troops.[179] Cromwell expounded the need to have an ideologically motivated force possessing the élan to carry the fight to the Royalists. He disciplined the Army, forming them into what he termed the New Model Army. To this end, the New Model Army was recruited mainly from Puritan adherents and included a large proportion of Leveller members and sympathizers.

The Cavaliers, still employing feudal organization and tactics, were no match for Cromwell's highly disciplined army. Hoping to escape the Roundheads, the King surrendered to Scottish forces in 1646. The Scots, however, sold him to Parliament in return for establishing Presbyterianism as the national religion. The King continued to plot, encouraging rebellions in Ireland where a Catholic army was killing Protestants. The final straw was Charles' incitement of Scottish Royalists to invade England. Cromwell defeated the Scottish army at the battle of Preston in 1648; in the aftermath the King was put on trial for treason.

Cromwell's Coup And Tyranny: The Protectorate

Cromwell had been spying on the Levellers.[180] He controlled the Army; High-ranking officers were allies and often shared family ties, particularly the highly influential Henry Ireton, his Son-in-Law. Parliament had still been trying to reach an accommodation with the King. To prevent this, Cromwell and his army occupied London. Now a Colonel in the New Model Army, John Lilburne had been against the occupation but was bought off with assurances the Leveller constitutional agenda would be enacted.

Members of Parliament favoring an agreement with the King were removed in "Prides Purge". Colonel Pride, commanding a large contingent of the New Model Army, surrounded the Parliamentary Hall preventing "disloyal" members from entering. The members who remained became known as the "rump parliament." Only then was the King's trial held. By one vote Charles was found guilty of treason against Parliament and the people it represented; he was executed in 1649.

The King's execution prompted Loyalist uprisings in Ireland and Scotland. Cromwell brutally suppressed the Irish and decisively defeated the Scots. The Levellers in the Army mutinied in opposition to both the Irish campaign and

---

[179] Rees, *The Leveller Revolution,* 76.
[180] Ibid, 56.

to seeing their political agenda sidelined. They were defeated, the Leveller leaders were imprisoned or executed and purged from the Army. With his military enemies defeated, his parliamentary enemies expelled, and the Levellers purged from the Army and maneuvered into political irrelevancy, Cromwell declared himself Lord Protector. He held this title until he died in 1658.

Democratic Transition: Counter Revolution And Glorious Revolution

Cromwell left power to his son Richard who proved weak and ineffective. Parliament called on Charles's son to return and claim the throne. Charles II and his brother James II had been in exile in France. There they learned the power of absolute Monarchy from the "Sun King" Louis XIV of France. The Pope and the Catholic Church justified the concept of divine right that God appoints kings conferring the right to rule absolutely. Absolute rule included the right, among others, to impose taxation. However, Parliament claimed the right to tax as a means to constrain the power of the King.

As a term of his coronation, Charles II agreed to remain Protestant and that the imposition of taxation would require Parliament's approval. To circumvent Parliament, Charles made an agreement with France to convert to Catholicism and join in a war against Protestant Holland. The war, however, was lost, and Charles incurred increased debt. He nevertheless maintained his Crown and, despite the opposition of Parliament, which he dissolved, succeeded in securing the succession to the throne by his Catholic brother James. Those who supported James' exclusion from succession formed the Whig party; those who supported his succession formed the Tory party. When James II succeeded to the Crown, he moved quickly to secure his position by enlarging the standing army, officered increasingly by Catholics, and packing Parliament with loyalists. The Whigs and Tories, seeing Parliament's loss of power, united to support Mary Stuart, the Protestant daughter of James II, and her husband William of Orange, the Dutch Republic's military commander.

James was planning to support France in yet another war against Holland. In 1688, William accepted the Whigs and Tories' invitation to take the throne and invaded England. Much of the King's Army defected, and James fled to France. James' daughter Mary was crowned but died within a few years, leaving William the King of England. In exchange for Parliament's support to the Dutch in their war against France, William and Mary accepted Parliament's new constitutional settlement.

Expansion and Export

During this period, the question of taxation was temporarily resolved, and in it, we see the origins of the American problem with taxation without representation. English Kings no longer maintained a lifelong authority to levy taxes. Rather Parliament reauthorized taxes regularly and began to specify how taxes could be used. Eventually, Parliament took much of the King's financial powers for themselves.

Both before and after the Civil War, English dissidents fled persecution by settling in North America. Massachusetts, in particular, was settled by staunch Puritans. During the War, some colonists returned to fight in the Parliamentarian Army. After the War, those fleeing both Cromwell's dictatorship and the restored Monarchy found sanctuary in the New World. The refugees passed on their ideals of religious freedom and republicanism to their children. A century later, those descendants became American Revolutionaries.

Summary Of The English Revolution

The English Revolution began as a power struggle between the nobility and the King. As the base of those who were politically invested expanded, the House of Commons eventually became preeminent over both the House of Lords and the Monarchy. While there was politically motivated violence, in the form of assassinations and plots, clandestine terrorist cells as we know them today did not exist. Clandestine networks operating at this time were not formed to commit political acts of violence. These networks were printing and distributing illegal literature and conducting religious services in the face of persecution. The English Civil War in the 1640s resulted in the beheading of King Charles I and Oliver Cromwell's reign until 1659.

Much of this era's violence was exacerbated because many Englishmen felt threatened by Catholicism, either French or Spanish. This fear increased its radicalism, allowing extremes such as the war of extermination against Catholics waged by Cromwell in Ireland. With Cromwell's death, Charles's son restored the Monarchy, but the historical dispute remained unresolved. In 1688 men opposed to King James II conspired to overthrow him with the help of the Kings' daughter Mary and her husband William of Orange, the ruler of the Dutch Republic.

The Glorious Revolution of 1688 has been called the first modern revolution.[181] Its modernity is constituted by the political regimes founded on freedom and equality, based on the consent of the governed. King William of Orange, the ancestor of George III, gained England's throne when James the II was deposed in 1688. James was an absolute monarchist, while William, the Prince of the Dutch Republic, was a product of the Reformation. William allowed a new constitution and more powers for the Parliament.

The English Civil War and Revolution of 1688 are, in many ways, the father of the American Revolution. The English writers who emerged from this period, such as Thomas Hobbes (1588-1679), John Locke (1632-1701), and

---

[181] Michael D. Richards, *Revolutions in World history*, (New York: Routledge, 2004), 11.

David Hume (1711-1776), had a significant influence on the American revolutionaries. Such writers justified resistance to unjust rulers and at the same time expressed the need for restraint, warning against the evils of chaos and disorder and, therefore, the need for authority and, most importantly, the idea of a social contract between Government and the governed. Good Government is desirable, and authorities are given their power by God or, in the secular view, granted by the people. Only when rulers become corrupt and misuse the power entrusted to them are the people justified and even compelled to revolt. According to Hume, in all governments, there is a perpetual struggle between authority and liberty.[182]

THE AMERICAN REVOLUTION

Rebellion to tyrants is obedience to God.[183]

The American Revolution did not begin as a planned revolt. Instead, it evolved into one due to State repression. However, the Americans did organize a clandestine resistance organization and later a shadow government.

Ideological development

There is some debate as to what the American Revolution was all about. Pauline Maier laid out three primary lines of thought: First, it was a social revolution, a class struggle, the worker/proletariat Americans throwing off the King, Second, it was a product of the Great Awakening and founded on Christian faith, Third it was a product of the Enlightenment and specifically of English radical thought expressed by men like John Locke. [184] That the American Revolution did not culminate in tyranny is, at least in part, because it is a direct descendant of the Protestant Reformation by way of the English Revolution of the 17th Century.

American Colonial resistance to English rule did not begin as a bid for independence; the Americans were committed to their British citizenship. The American colonists saw themselves as upholding the traditional rights of Englishmen. Colonial leaders followed in the footsteps of the radicals of the English Civil War, such as the Levellers, who had called for representative Government, an end to absolute Monarchy, advocated for civil rights, freedom from quartering troop, and freedom of the press.

---

[182] David Hume, *"Of the Origin of Government,"* in *Essays Moral, Political, Literary*, edited by Eugene F. Miller.

[183] Attributed to Benjamin Franklin: Michael Medved, *The American Miracle: Divine Providence in the Rise of the Republic,* (New York: Crown Forum, 2016), 284.

[184] Pauline Maier, *From Resistance To Revolution: Colonial Radicals Development of American Opposition To Britian, 1765-1776* (New York: W.W. Norton & Company, 1991), vi & xi.

One of the most influential thinkers of the Enlightenment on the American Revolution, John Lock wrote during the 17th Century.[185] He had gone into exile during the English Restoration (1660-1688) and helped write the Constitution of Carolina in 1669. Locke argued that people are born with certain natural rights; they can give up some of their rights in exchange for the security offered by the State. If the State fails to provide protection, then the people have the right to exchange the Government for a better one.

Following the Seven Years War, known in the colonies as the French and Indian War, England was heavily in debt. Among steps taken by the English Government to address its financial difficulties was the Stamp Act of 1765. The act placed taxes on intended legal, commercial, and other documents in the colonies. The colonists objected to the imposition of this tax without their representation in Parliament. The Colonies boycotted English goods and succeeded in forcing the repeal of the Stamp Act.

When George III ascended to the throne, it soon appeared to the colonists that he was attempting to reinstate absolute Monarchy. Men such as John Adams began to organize and speak out and delineate what constituted lawful and unlawful violence in defense of colonial rights. Adams echoed Cicero of ancient Rome, who stated, "the people's good is the highest law," saying the Government's purpose was "the greatest quantity of human happiness."[186]

The Colonists had many political allies in England. John Wilkes, Minister of Parliament for Middlesex County and later the Lord Mayor of London, kept regular correspondence and coordinated efforts with the colonial resistance leaders. Wilkes was waging political battles in Parliament to uphold the same rights the colonists were fighting for.

Throughout their resistance to the Stamp Act and later Acts, Colonial leaders did not condone violence rather, they took steps to maintain order during the resistance. Furthermore, the leaders stressed that what disorder did occur was an indictment of the Government and proof of poor governance. The Colonists asserted uprisings were only possible in a free society. Under a tyranny, the people are forced into submission. The way to prevent uprisings was not repression but reform.

Thomas Paine, an English radical, arrived in America in 1775. He became enormously influential with his pamphlet *Common Sense.* He argued for independence from Britain and laid out the case for America's military capability

---

[185] Brenée Goforth Swanzy, "How John Locke Influenced the Declaration of Independence," John Locke Foundation, March 27, 2022.

[186] Pauline Maier, *From Resistance to Revolution: Colonial Radicals and the Development of American Opposition to Britain, 1765-1776* (New York: W.W. Norton & Company, 1991), chapters 2-4.

to gain it.[187] Later, after the Continental Army fled New York in disarray, General Washington had Paine's pamphlet *The American Crisis* read to steel his soldiers' resolve.[188]

The Vanguard

Before the outbreak of war, there were remarkably few deaths caused by small group action and civil unrest. This is because of American ideology, which did not condone illegitimate violence. The clandestine organization, the Son of Liberty, ensured violence was regulated and applied judiciously; their organization fused Small group Action and Mass Civil Unrest. The Sons of Liberty organized to resist the Stamp Act in 1765, and with its repeal, the Sons of Liberty disbanded only to reform anew in the 1770s. Their most well-known action was the Boston Tea party carried out in 1773.

Isaac Barre, another Member of Parliament and ally of the Colonists coined the title Sons of Liberty. Barre said Americans were loyal subjects protecting their liberties against policies that made "the blood of these sons of liberty to recoil within them."[189] The Sons of Liberty enforced conformity to colonial boycotts of British goods and dissuaded customs officials from carrying out their duties through coercion and threat of violence. Actual violence was, however, quite rare.

The English Parliament responded to the Tea Party with repressive measures: town meetings required the Governor's approval, the power to appoint officials was taken from local Government and given to the Governor of Massachusetts, and Boston harbor was closed. General Thomas Gage was named Governor and given authority to quarter troops in private homes. Boston was effectively under martial law and ordered to repay the lost tea, equivalent to some 1.5 million dollars. The stage was set for armed conflict.

Another lesser-known group was the Regulators of North Carolina. The Regulators were formed to control banditry in the far west of the colony. The Regulators began an armed insurgency against the Colonial government. New England's hailed the regulators as justly rising up when the colonial Government gave no recourse. Local Sons of Liberty, however, sided with the Colonial Governor; they knew he was about to grant their petitions and so felt the Regulator uprising was unjustified. In fact, the leader of the Cape Fear Sons marched with the Governor to the Battle of Alamance, where the Regulators were defeated.[190]

---

187 Thomas Paine, *Common Sense and Other Writings*, (New York: Barnes & Noble, 2005), 17-69.

188 Ibid., preface.

189 Maier, *From Resistance to Revolution*, 82.

190 Maier, *From Resistance to Revolution*, 86 & 197.

Mass Civil Unrest And Revolution

This stage in the Colonies was an interesting mix of legal posses and extra-legal mobs. Several times, organized crowds acting responsibly were followed by riots pursued for personal reasons of gain or revenge. Resistance leaders condemned the riots and went to great lengths to ensure later actions did not descend into chaos. Recognizing and guarding the individual rights of those they opposed is a distinguishing mark of the American Revolutionaries. Additionally, the protest leaders were often municipal officials acting to protect the local populace, such as the actions taken to counter the impressment of colonial sailors by the British Navy.

At a time when there was no established police, posses were formed to enforce the law and arrest criminals. Extra-legal mobs created mass disturbances. "Extra-Legal" because they quite often consisted of the same individuals who made up legal posses. When lawful authorities failed to act, these mobs acted to defend their communities' interests, punish outlaws, and prevent abuses by public magistrates. Mobs were a form of direct democracy.[191]

In Boston, a mob leader Known as "General" Mackintosh wore a uniform and had designated officers controlling the actions of the masses. Those under his "command" were issued instructions to maintain order and limit violence and property damage.[192] Only officers were permitted to carry weapons, in their case, sticks. Despite actions taken to mitigate violence, bloodshed sometimes resulted anyway.

Resistance to the 1767 Townsend Acts caused Massachusetts Governor Bernard to ask for troops to maintain order. Rather than order, this resulted in more violence when in 1770, rioters attacked an army barracks and the soldiers fired upon the rioters killing five. This became known as the Boston Massacre. Like Bloody Sunday in Russia, this incident became a rallying cry for the colonists. Colonial leaders demanded the troops' removal and threatened to raise 10,000 citizens to drive the soldiers out. The Governor complied, and Parliament further eased tensions with the repeal of the Townsend Act; a tax on tea, however, was left in place.[193] Other acts of mass violence included the tar and feathering of supporters of the Crown, who were sometimes beaten and strangled, but not to death.

Taxation was not the only grievance the Colonists had toward the English Government. Petty abuses by British officials, particularly naval officers, created a more significant rift between Colonists and the Crown. For example,

---

[191] Ibid, 100-120.

[192] Ibid., 69-70.

[193] Robert J. Allison, *The American Revolution: A Very Short Introduction*, (New York: Oxford University Press, 2015), 15.

Captain Reid of the *Liberty* was accused of acting "more like a pirate than one appointed or commissioned to protect our trade." Colonists sunk the ship in 1769. Riots against the British policy of impressment of American sailors resulted in the imprisonment of British Sailors. In 1767 the town watch of Norfolk, Virginia, called out the townspeople who captured and imprisoned a press-gang from the British warship *HMS Hornet,* and in 1764 the town of Newport fired on the Schooner *HMS St John.*[194]

The colonists particularly reviled Lieutenant Duddingston and his ship, the *HMS Gaspee.* Duddingston had regularly provisioned his vessel through the confiscation of private property from local farmers, and his crew beat colonists suspected of smuggling and stole their cargo. The *Gaspee*'s activities drove up the price of food commodities. The Admiralty ignored complaints from the Governor of Rhode Island. In 1772 when the *Gaspee* ran aground, Rhode Islanders boarded and burned her. The British Government demanded the perpetrators be arrested and sent to England to stand trial. The Colonial Government refused to comply; the colonists felt justified because Duddingston had acted illegally, and the British Government failed to redress the problem.[195]

Following the Tea Party in 1773, England, believing the other colonies loyal, moved to isolate Massachusetts in hopes of preventing the spread of sedition. However, the Sons of Liberty in Boston had already established a communications network with Sons of Liberty organizations in the other colonies. Samuel Adams also set up the Boston Committee of Correspondence, linking the city with the surrounding communities.

Rhode Island called for the Assembly of a Continental Congress. Every colony except Georgia was represented when they met in 1774. The Congress petitioned the King to end Boston's isolation and change specific policies deemed against the English Constitution, the quartering of troops among them. The Congress resolved to meet again the following year.

Meanwhile, General Gage, in an attempt to maintain order, consolidated gunpowder in Boston, gunpowder that had been in provincial storehouses. Gage feared radicals would steal the munitions. His actions raised fears among the colonists who began forcing loyalists to flee their homes and seek British protection in Boston. In New Hampshire, a local militia overwhelmed a small British garrison confiscating its canon and munitions. The Americans blocked road passage to British soldiers but never harmed them. Former Prime Minister William Pitt suggested pulling the troops out and reducing tensions. Instead, the Government doubled down, declaring

---

194 Maier, *From Resistance to Revolution*, 75-90
195 Ibid, 6-12.

Massachusetts in rebellion, ordering the arrest of provincial leaders, and authorizing military force. General Gage then marched on Lexington and Concord with 800 men, where the shot was heard around the world.

By the time the Continental Congress reconvened, New England had rallied to Boston's defense, and the Vermont Militia had captured Fort Ticonderoga, whose cannons would later be carried to the siege of Boston by Henry Knox. Congress adopted the New England militia as the Continental Army and appointed George Washington as its commander. Still, Congress stated its loyalty to the Crown and asked the King to redress colonial grievances. By the third Continental Congress, it was clear the King would not bend. According to the Colonists' ideals, he acted unjustly and was therefore no longer a legitimate ruler. The people were free to rule themselves.

Despite common views to the contrary, much of the war was waged according to the standards of European warfare. Washington maneuvered his army against the British military, and battles were fought with lines of men firing muskets at one another. The War in the South was different. Here there was Terror. Civilians, both Rebels, and Loyalists, were targeted with violence and intimidation; reciprocal violence between rebels and loyalists occurred regularly.

Civil War

The American revolution was in many ways a continuation of the English Civil War. The colonists were Englishmen fighting other Englishmen. But with the declaration of independence the Colonists declared themselves a sovereign nation making the conflict a war between nations. However, patriots supporting independence were in a civil war against colonial loyalists who fought for the king.

Revolutionary Victory And Consolidation Of Power

In this stage, the radicals should have eliminated the moderates and seized power and then continued to fight amongst themselves for leadership. That did not happen. Instead, the Continental Congress created an equitable, albeit imperfect, system of governance. With the exception of 1860-65, the US has been peacefully transitioning power between political parties for over two hundred years.

The American desire for order and good governance led to many compromises as Congress worked out the policies of the newly established United States. Slavery was a tragic victim of the need to avoid instability. The reckoning was postponed for another 70 years.

Coup d'etat

Violent radicals never succeeded in co-opting the Revolution. General Washington ensured the Army did not mutiny. Because the States were jealous of their autonomy, it would have been challenging to gain power over the entire country without the Army. The new republic did not face a large hostile military force on its border. Canada to the North posed a threat, but that threat required Britain to project power across the Atlantic. This did happen in 1814 but was never a significant enough threat to cause internal suspicions and panic, which could increase the position of radicals.

Tyranny

Following the war, King George III asked if General Washington would use his army to seize power. He was told that, in all likelihood, Washington would retire to his home in Virginia. To which the King reportedly replied that if he did so, Washington would be the greatest man in history.[196] George III had good reason to ask. The history of the English Civil War offered a clear precedent. When Charles I lost to Parliament's forces, he was tried and beheaded in 1649, and the leader of the Parliamentary Army, Oliver Cromwell, dissolved Parliament and declared himself Lord Protector, ruling England as a military dictator for nearly a decade before the monarchy was restored under Charles II in 1660. The pattern was familiar: revolutionary armies, having shed blood for a cause, rarely surrendered power voluntarily. Cromwell's rule demonstrated that even a revolution fought in the name of parliamentary liberty could produce a tyranny worse than the one it replaced.

Washington faced his own version of this temptation. When Congress proved unable to pay the Continental Army, discontent among the officer corps reached a breaking point. In March 1783, a group of officers at the army's encampment in Newburgh, New York, circulated anonymous letters urging the army to refuse to disband until Congress met its obligations, with the implied threat of a military takeover of the government. Washington intervened personally, addressing his officers and appealing to their honor and to the principles for which they had fought. In a moment that has become legendary, he paused to put on his reading glasses, remarking that he had grown gray in the service of his country and now found himself growing blind as well. The gesture disarmed the conspiracy. Washington would countenance no seizure of power, no matter how justified the grievances behind it.

One man's moral restraint made a difference that shaped the course of the next two centuries. Washington's voluntary relinquishment of military authority established the precedent of civilian control that became foundational to the American constitutional order. That the United States did not descend into tyranny and terror was also due, in part, to its geographical remoteness. Separated from European great-power rivalries by the Atlantic Ocean, the young republic faced no immediate external threat that could justify the concentration of military power in a single leader's hands, removing one of the

---

[196] Allison, *The American Revolution*, 85.

key pressures that had driven revolutionary France and Cromwell's England toward dictatorship.

Export And Expansion

Some will say the treatment of Native Americans is tantamount to terrorism. This may have been true in some cases, such as the Trail of Tears initiated by Andrew Jackson. Westward expansion had been one of the points of contention between the Crown and colonies. Britain had attempted to restrict westward expansion as they wished to avoid further troubles with the natives. It was this that had triggered the French and Indian War. With independence, the Americans expanded relentlessly toward the Pacific coast.

It was, however, the idea of Government by the people which Americans most desired to export. Republican ideas were heralded as beneficial for all men in all nations. America does not justify illegitimate violence; it stresses adherence to the rule of law and the right to self-government. Yet, promoting democracy has been a national goal for the Department of State.[197] Former Secretary John Kerry stated,

> *Strong civil society and respect for democracy, the rule of law, and human rights – these are not just American values. They're universal values. They're universal aspirations.*[198]

Problems arise when Americans promote representative democracy internationally through military force, as it tried to do in Afghanistan and Iraq. Such adventures are in direct opposition to the ideals of popular sovereignty. A government of, by, and for the people means the people of that particular nation. Foreign intercourse yes, foreign entanglements no.

The American Revolution represents the RTVC's most challenging case. Unlike the French, Russian, or Iranian revolutions, the American movement was led not by ideological vanguards seeking to destroy the existing social order but by colonial elites seeking to preserve rights they believed they already possessed. The absence of a utopian ideological program, the presence of institutional continuity through colonial legislatures, and the moral restraint of key leaders (Washington's refusal of kingship, the Constitutional Convention's willingness to compromise) meant that the conditions producing the RTVC's tragic arc were largely absent. This suggests that the RTVC's strongest explanatory power applies to revolutions driven by utopian ideology and vanguard organization rather than to all revolutionary upheavals. The American

---

[197] Marian L. Lawson and Susan B. Epstein, *Democracy Promotion: An Objective of U.S. Foreign Assistance*, Congressional Research Service Report R44858, updated January 4, 2019.

[198] U.S. Department of State, "Promoting Resilient, Open, Democratic Societies," last modified [date if known or n.d.], accessed October 18, 2025.

case is less a confirmation of the model than a demonstration of the conditions under which the model's logic can be averted.

THE FRENCH REVOLUTION

Like Saturn, the Revolution devours its children.
-Jacques Mallet du Pan

The French Revolution was, in many ways, spontaneous. Rather than a planned revolution, men took advantage of situations as they were presented. The National Assembly became more radical to appease the masses, who became more unruly as the Assembly indulged them. This radicalism devolved into the Terror. Eventually, Napoleon brought enforced sobriety to the country.

Besides the Enlightenment, the other significant influence on the revolution was the example of the Americans. Lafayette and other officers returned from service with General Washington and called for a constitutional monarchy. Thomas Jefferson was the American envoy to France and Thomas Paine, whose pamphlets were so influential in America, became a member of the French Assembly. Paine also influenced *The Declaration of the Rights of Man and of Citizens by the National Assembly of France;* one of the stated rights is the right to resist oppression.[199]

18th Century France was the most populous nation in Europe and at the center of Western Civilization. Like Czar Alexander II a century later, Louis the XIV attempted to modernize a country still largely mired in a feudal system. France faced an Economic crisis caused by, among other things, war debt. The Government was the chaotic result of hundreds of years of Monarchy with an ad hoc system of administration. Louis attempted to reform the bureaucracy to more efficiently administer his government and collect taxes. This attempted centralization of government resulted in disaffection of a number of the aristocracy and unrest among the larger population.

Ideological Development

Enlightenment thinkers such as Jean Rousseau and Voltaire were very influential in setting the stage for the Revolution. Rousseau advocated rule bythe "general will" of the people and Voltaire ridiculed the privileges of the aristocracy and the clergy. Yet neither advocated terrorism; indeed, Voltaire became a part of the establishment and may be considered a monarchist.[200]

---

199 Paine, *Common Sense and Other Writings*, xxvii-xxx; Allison, *The American Revolution*, 99; French National Assembly, "The Declaration of the Rights of Man and of the Citizen," Élysée, last modified December 14, 2022.

200 Pierre Brodin et al, *La Revolution Francaise, La France D'Hier et D'Aujourd'hui* (New York: The Gordon Company, 1938), 10.

Nevertheless, the ideas these men promoted undermined the self-confidence of the elites.

Rousseau's *Social Contract,* 1762, argued that Government existed to serve the people rather than the people existing to serve the Government, or rather no one had a right to rule by might, as this is not moral. It is only through mutual agreement or contract that a people consent to be governed. Rousseau was born in Geneva, a self-governing theocracy established in the 16th Century when the Reformist Jean Calvin was influential.[201] But he became *persona non grata* for both his political and anti-Christian views. Rousseau said Christianity had been responsible for "the most violent of earthly despotisms" based on lies and error. He paved the way for the anti-clerical violence of the revolution.[202]

Under the French *ancien régime*, society was divided into three legally distinct orders. The First Estate comprised the clergy, from bishops and abbots to parish priests, who controlled vast landholdings and were largely exempt from taxation. The Second Estate consisted of the nobility, who held privileged access to military commands, high offices, and feudal rights over land and labor, and who likewise enjoyed significant tax exemptions. The Third Estate encompassed everyone else, from wealthy bourgeois merchants, lawyers, and professionals to urban laborers and the vast peasant majority, who together constituted roughly 97 percent of the population and bore the overwhelming burden of taxation. This arrangement, formalized through the Estates-General, ensured that the two privileged orders could always outvote the Third Estate, preserving an aristocratic social order that concentrated privilege at the top while distributing its costs downward.

Some Clerics were advocating for change as well. Abby Sieyes wrote the influential pamphlet, *What is the Third Estate?*[203] In it, he asserted the third estate was nothing short of the nation. The nobles, he wrote, were not of the people; their interests were personal and not the nation's interests. The nobles were, in effect, foreigners.

Where the elites of the Old Regime accepted Voltaire and the Grand Philosophes, Sieyes made a direct attack on the privileges of the aristocracy. Writers unable to break into gentile society began writing *Libelles* in a literary underground.[204]

---

[201] Brodin, *La Revolution Francaise*, 10.

[202] Jean-Jacques Rousseau, *The Social Contract and Discourses*, trans. G. D. H. Cole (London: J. M. Dent & Sons, 1913), bk. 4, ch. 8. available at The Social Contract & Discourses by Jean-Jacques Rousseau | Project Gutenberg.

[203] Ian Davidson, *The French Revolution; From Enlightenment to Tyranny*, (New York: Pegasus Books, 2016), 15.

[204] Robert Darnton, *The Literary Underground of the Old Regime*, (Cambridge, MA:

> *It was in the depths of the intellectual underworld that these men became revolutionaries and that the Jacobinical determination to wipe out the Aristocracy of the mind was born.*[205]

Resenting their inability to rise from the confines of their social position, men such as Jean-Paul Marat and Jacques-Pierre Brissot attacked the establishment in libelous pamphlets.

With revolutionary success, some of these men became increasingly vicious in their writing. In July 1790, warning of the threat of a counterrevolution, Jean-Paul Marat wrote,

> *Five or six hundred heads cut off would have assured your repose, freedom, and happiness. A false humanity has held your arms and suspended your blows; because of this millions of your brothers will lose their lives.*[206]

The Sans-Culottes, literally "without breeches," were the radical working-class populace of Paris, so named because they wore the long trousers of common laborers rather than the knee breeches of the aristocracy. Among the most extreme factions were the *Enragés*, or "enraged ones," a group of radical agitators who demanded price controls, harsher measures against hoarders, and immediate relief for the urban poor. The *Enragés*' leader, Jacques Roux, laid out the *Manifesto of the Enragés* in 1793, in which he threatened the revolutionary National Convention to heed the demands of the underclasses of Paris.

> *We know that no doubt there are evils that are inseparable from a great revolution…we also know that the people have been betrayed by two legislatures..and that its time that the sans-culottes who smashed the scepter of kings see the end to insurrections and all types of tyranny.*[207]

The base passions of the Terror soon overshadowed the grand ideals of the Enlightenment.

The Vanguard

Radical Clubs such as the Jacobins and Girondins maybe the closest to small groups found in the French experience. The Jacobins were a radical political faction that took their name from their meeting place, a former Dominican monastery of Saint-Jacques in Paris. Initially a broad coalition of revolutionary reformers, the Jacobins grew increasingly radical under the leadership of Maximilien Robespierre, eventually championing centralized

---

Harvard University Press, 1982)

[205] Darnton, *The Literary Underground*, 21.

[206] Jean Vidalenc, "Jean-Paul Marat: French Politician, Physician, and Journalist," Encyclopedia Britannica.

[207] Jacques Roux, "Manifesto of the Enragés, 1793," Trans. By Mitchell Abidor, Marxist Internet Archive, accessed 3 June 2017.

authority, direct democracy, and the violent suppression of enemies of the Revolution. They drew much of their support from the Parisian populace and dominated the Committee of Public Safety during the Reign of Terror.

The Girondins, named for the Gironde department in southwestern France from which several of their leading members hailed, represented a more moderate republican faction within the National Convention. They favored a decentralized government, feared the growing influence of the Parisian mob, and advocated for extending the Revolution through war abroad. Though they shared the Jacobins' commitment to a republic, they resisted the concentration of power in Paris and opposed the extreme measures that Robespierre and his allies pursued. The rivalry between these two factions defined much of the Revolution's internal politics until the Jacobins purged the Girondins in 1793, sending many of their leaders to the guillotine.

Unlike the Sons of Liberty, the French clubs were not clandestine and did not engage in small group actions. The Jacobin club did, however, resort to Terror, as we shall see below.[208]

While there were no clandestine terror cells, there were underground networks. As in the English and American revolutions, pamphlets were a primary means of disseminating revolutionary ideas. The Royal Government heavily censored the press and maintained tight controls on imported literature. As a result of the State's repressive censorship, rather than diminish, a vibrant underground for literature developed to disseminate revolutionary ideas. Clandestine networks developed for printing and distribution as well as for smuggling censored material into the country. Some of this developed during the Reformation, as Protestants attempted to make French-language Bibles and other Protestant literature available. Literature critical of the Monarchy was censored under Louis XIV. Voltaire was a victim of this censorship and found himself jailed or exiled on numerous occasions.

As the Girondin Antoine Barnave said, "those making revolutions don't do so with metaphysical maxims…the masses are not mobilized, except by realities."[209] In other words, it was not the ideas of Rousseau but the socio-economic hardships that made the revolution happen. Initially, there was no thought of doing away with the Monarchy; the Revolution began as a reform movement, and it quickly spiraled out of control. Like Czar Nicolas, Louis XIV was a weak ruler, and as in the Russian Revolution, the French Revolution resulted in a Dictatorship. While not calling for violence, the result was anti-clerical violence and mob rule.

---

[208] The Editors of Encyclopedia Britannica, "Jacobin Club," Encyclopedia Britannica, accessed June 3, 2017.

[209] Brodin, *La Revolution Francaise*, 10.

REVOLUTIONS and CONSTITUTIONS

Mass Civil Unrest And Revolution

Hoping to find a solution to France's financial problems, the King called for a general assembly of the three estates: nobles, clergy, and commoners. In the Old Regime, commoners lacked room for upward mobility except through the judicial arm; this meant many in the third estate were educated lawyers. The third estate outnumbered the other two estates and wanted their numbers recognized. Traditionally the estates met and voted separately, which meant the first two would always outvote the third. The Third Estate forced the creation of the National Assembly, a rogue Parliament, where dissidents from the first two estates eventually joined them.

The King was forced to recognize the National Assembly, but seeing power slipping from his control, he moved troops closer to Paris. The National Assembly, in turn, feeling threatened, called for popular support. The result was the storming of the Bastille by the Sans-Culottes and the creation of a "National Guard" under the Assembly's control. Several months later, a mob of 6000 women and the National Guard marched on the King's palace in Versailles. The Mob killed Several of the King's guards and returned him to Paris as a prisoner.

Revolutionary Victory And Consolidation Of Power

Over the next two years, with the King in house arrest, the Assembly restored order. By 1791 the Assembly had reorganized France as a constitutional Monarchy. The King, however, attempted to escape the country. Suspecting Louis was conspiring to return with an army to restore his absolute Monarchy raised tensions among the lower classes, who were also discontented with the Assembly. The Middle Class, the Bourgeoisie, dominated the Assembly; the lower classes felt they were underrepresented.

The radical Bourgeoisie factions, the Jacobins and Girondins, called for the end of the Monarchy and the establishment of a republic. In the spirit of Rousseau, the Jacobins called for equality for all men, thus winning the support of the mass of lower-class Parisians. Meanwhile, foreign Monarchs, among them Queen Marie Antoinette's Brother Leopold of Austria, called for the Restoration of the French Monarch. French men and women responded in the thousands to the call to war. Initial poor showing on the battlefield undermined the Assembly. The Jacobins led by Georges-Jacques Danton, backed by the Paris commune's mobs, seized power in the Assembly and executed several thousand of their Girondin opponents.

The Jacobins dissolved the Assembly and established the National Convention with members elected by universal male suffrage. Maximilian Robespierre shared leadership of the Jacobins with Danton, both aided by Jean-Paul Marat, who published a radical newspaper used by the Jacobins to denounce the revolution's enemies. The King was tried for treason and, like

Charles I of England, was found guilty by one vote and sentenced to death.

Led by Jacques Brissot, the Girondists, whose power base was in the countryside and cities other than Paris, staged an uprising against the Jacobin government. Once again, the Jacobins called in the Sans-Culottes, who stormed the National Assembly and removed the Girondists. When a Girondists murdered Marat, Robespierre used the anti-Girondists sentiment to seize greater control. Under Robespierre, the Jacobins moved quickly to suppress the Girondists, sending Brissot and one thousand more to the guillotine, initiating a year of Terror. With the Terror, the Jacobins reorganized the Government and the Army and successfully drove out the invading foreign armies and crushed counterrevolutionary uprisings, such as that in the Vendee.

Civil War

The French remained loyal to the Catholic Church during the Reformation. Arguably, Protestants ceased to be a significant factor in France following the St. Bartholomew's day massacre in 1572. When it did reject its Catholic background, rather than become Protestant, it became Secular Humanist and often violently anti-clerical.

The anti-clerical sentiments of the revolutionaries led to purges of clergy throughout the nation. In the Vendee, a province in west-central France where the revolution was not popular, the clergy were instrumental in leading a counterrevolution. The Revolutionary Government raised 12 army divisions known as the Infernal Columns and waged a three-year war of extermination resulting in between 60,000 and 200,000 deaths.[210] Fear of counter revolutionary forces both real and imagined led to ever greater excesses as the Radicals exterminated any and all who were deeded a threat.

Robespierre and the Jacobins allied with the Sans-Culottes masses to impose the Terror to gain and maintain power. The Jacobins formed the Committee of Public Safety headed first by Danton and then by Robespierre. The CPS enforced conformity to the revolution, as defined by the Jacobins, through Terror and intimidation. Counterrevolutionaries and reactionaries were tried and guillotined during the infamous Reign of Terror from 1793 to 1794. Eventually, along with some 16,000 others, both Danton and Robespierre succumbed to the Terror they had created.

Coup D'etats and Tyranny

Robespierre consolidated power for the Jacobins and then purged his rival Danton, sending him to the guillotine. Robespierre had succeeded in

[210] J. Llewellyn and S. Thompson, "The Vendee uprising", Alpha History, accessed June 3, 2017, http://alphahistory.com/frenchrevolution/vendee-uprising/.

consolidating power for the radicals and was the dictator of France for a short time. He was unable to maintain control and was himself guillotined. After his death, the Government reorganized and dissolved the CPS and the National Convention, creating the Directory in 1795. The new Government consisted of an executive body of five directors and a bi-cameral legislature. However, popular discontent resulted in rioting by mobs attempting to storm the Government. This time, however, a ruthless revolutionary General was called upon to protect the Government. Napoleon cleared the streets with canister shot from his canons, the famous "whiff of grapeshot." In 1799 He helped overthrow the Directory and establish the Consulate. With control of the army in his hands, Napoleon quickly named himself First Consul; by 1804, he was secure enough to crown himself Emperor.

Export and expansion

The Revolutionary Government exported its ideology first through ideas and then through conquest and the imposition of revolutionary states. Even before Napoleon, French Revolutionary Armies spread their ideals of liberty and "liberated" their neighbors. The Napoleonic wars stretched the borders of France to Moscow, Italy, and Spain. Ultimately Napoleon was defeated, and the Monarchy was restored. But Louis XVIII's reign lasted only a short time. France saw revolution again in 1830 and 1848. The French Revolution inspired popular uprisings throughout Europe during the 19th Century. Russian Revolutionaries also looked to the French for inspiration, naming a battleship for Jean-Paul Marat.[211]

CONCLUSION

Political theory had not evolved to the point of planning Terror for a revolutionary cause until the middle of the 19th Century. The republican revolutions discussed above do not adhere neatly to the RTVC. The revolutions were largely unanticipated events emerging from the crises of the times. By contrast, the Russian and Iranian revolutions were calculated, or at least some had calculated and tried to provoke the unrest that led to revolution.

At the time, the French Revolution was, arguably, a much more significant event than the Revolution in the British colonies. The colonies were a backwater mainly of value in logistical support to the British colonies in the West Indies. France, on the other hand, was one of the two great world powers. It was one of the oldest civilizations, and its' language had been used as the

---

[211] Vidalenc, "Jean-Paul Marat."

standard form of diplomatic and commercial communication, so much so that the term Lingua Francia is used to this day.

As in America, the revolution the French fought was for freedom and equality. But, as Edmund Burke wrote in his *Reflections on the Revolution in France,* it was quite different from the English and American Revolutions. The former was conducted to preserve liberty and society from oppressive government seeking to impose tyranny. In contrast, the French Revolution sought to build freedom through the complete destruction of the *Ancien Regime.* While the British and Americans sought to preserve their heritage, the French struck out on a wholly new course of rationalism that resulted in anarchy, and as Burke predicted, in tyranny and a military dictatorship.

The French Revolution was influenced by ardent secularists such as Rousseau, Voltaire, and Robespierre. Some of the greatest excesses of the Terror were against the clergy and during the catholic counterrevolution in 1793. Men like Danton and Robespierre did not seem to be restrained by religious-moral values. In England, denominational zealotry resulted in backlash and led to ideas of religious liberties that found fruition in the American colonies. The American Revolution drew upon Calvin, primarily through the Puritans of New England, and benefited from Englishmen writing during and after the 17th-century upheaval in Britain. As a product of the English experience, the moral convictions of George Washington set a precedent for the peaceful transfer of power still enjoyed by the United States to this day.

If the English Revolution was the father of the American, the French Revolution was the father of the Russian. From the French Revolution, we get the political notions of the Left and Right, the radical Jacobins being on the Left. The conservatives on the Right were those wishing to preserve the Monarchy and the Clergy. After the French Revolution, political theorists saw that it was possible to bring about revolutions intentionally. Some Socialists and Anarchists began to advocate using Terror and violence as a catalyst for change, and terrorists became regular actors on the political stage. By the late 19th and early 20th Century, revolutionaries were organizing people and deliberately using terror and mob violence towards the political end of overthrowing the existing socio-political order.

The French Revolution is the father of Revolutionary Terrorism. It is also the father of total war. Invaded by foreigners, the Revolutionary Government mobilized the nation and built a patriotic army of the people rather than mercenaries that filled the royal ranks. Interestingly the term guerrilla warfare, meaning *small war*, came into use among those resisting Napoleon's occupation of Spain. More positively, the Revolution produced the Declaration of the Rights of Man and the Napoleonic code, a legal system used to this day.

Allan Bloom's *The Closing of the American Mind* has a chapter entitled, *Two*

Revolutions and Two States of Nature.[212] The two revolutions are the American and French Revolutions. The two states of nature being the worldviews predominant in each instance. Bloom has the Englishman John Locke represent the American Revolution. His was a utilitarian view of nature, which was to be dominated and used to enrich humanity. The Frenchman Rousseau, on the other hand, promoted reverence for nature and saw mankind as a part of it. Locke felt man needed Government to keep him from his evil ways; Rousseau felt Government was an evil that corrupted man who was essentially good in his natural state.

The difference, says Bloom, can be summed up in the term Bourgeois. Europeans saw the Bourgeois as the failure of modernity, greedy, self-serving, and corrupt. Americans, instead, used the term middle class who they glorify as the triumph of modernity. Bloom called the American Revolution, "the greatest transformation of man's relations with his fellows and with nature ever affected."[213]

Perhaps America lacked the cultural baggage of the French Royal State. Perhaps geography allowed America to expand with little resistance, and the lack of threat from neighboring countries helped avoid the paranoia that led to the Terror in France. In the final analysis, America escaped the Revolutionary Terrorism Value Chain's march to Tyranny because George Washington refused to seize power, as a lesser man might have done.

---

[212] Allan Bloom, *The Closing of the American Mind* (New York: Simon and Schuster, 1987), 157–72.
[213] Ibid, 158.

CHAPTER 4

RECIPROCAL RADICALISM:
Street Fighting In Weimar America

You say you want a revolution. Well, you know. We all want to change the world.
-the Beatles

In 1979 the American communist party organized a "death to the Klan" march in Greensboro, NC.[214] A coalition of Klansmen and Neo-Nazis rallied to counter protest; this represented a turning point for the extreme right in America. The merger of the KKK with the Neo-Nazis was new. Many Klansmen had fought during WWII and considered Neo-Nazis to be anti-American. The Klan's pro-America anti-Nazi stance changed post-Vietnam.

The Vietnam War was central to the narratives of both the left and right but for different reasons. The Radical Right saw the War as proof that the government had been usurped by leftists and no longer served the American people. The Left saw the War as proof of America's imperialism and fascism.

Greensboro highlighted a change for the left as well. The left moved somewhat away from class struggle and towards the anti-fascist /anti-racist struggle. Greensboro may be considered a continuation of the Left-Right struggle imported from Europe, each side feeding off the other in an escalating spiral of violence

From Weimar Germany to modern America, radical ideological movements have often fed off one another, escalating social conflict in a pattern

---

[214] Jason Kops, "The Greensboro Massacre: A Challenge to Accepted Historical Interpretations," *Explorations: The Journal of Undergraduate Research and Creative Activities for the State of North Carolina* 7 (2012): 76–86, accessed November 1, 2025, https://docslib.org/doc/13286066/the-greensboro-massacre-a-challenge-to-accepted-historical-interpretations-jason-kops-east-carolina-university-faculty-mentor-wade-dudley-east-carolina-university.

known as reciprocal radicalization. In *Murder in Amsterdam*, Ian Buruma observed that polarized societies create the conditions in which extremists on opposing sides function as mirrors of one another, each using the other's provocations to justify its own radicalism, a dynamic in which escalation becomes self-sustaining.[215] This chapter traces the dynamic between radical left-wing and right-wing street movements in 21st-century America, drawing historical parallels with Germany's interwar period and the United States in the 1960s.

Applying the RTVC framework to contemporary America requires an important caveat. The United States has not experienced revolution. What the framework illuminates here is not a revolutionary situation but rather the *pre-conditions* and *early-stage dynamics* that, in other contexts, preceded revolutionary escalation. The analytical question is not whether America is undergoing revolution but whether recognizable patterns from Stages 1 through 3 of the RTVC are visible in embryonic form, and whether awareness of those patterns might help prevent escalation.

Weimar Germany's street battles between Communist and Nazi militias, along with the failure of liberal centrists, illustrate how polarization and reciprocal fear destroy institutional legitimacy.[216] From the 1960s through the early 1980's radical groups like the Weather Underground and far-right vigilantes escalated urban violence in America. The ideological descendants of these groups are visible today in Antifa, the Patriot Front, and other militias.

Reciprocal Radicalism And The Imagined Enemy

Reciprocal Radicalism is a term, adapted from a paper written by Benjamin Lee and Kim Knott, referring to the idea that radicals feed off their opponents and competitors.[217] Opponents are groups perceived as enemies and competitors are those sharing the same perspective but vying for leadership of the same movement. The resulting retaliatory attacks escalate in frequency and intensity.

---

[215] Ian Buruma, *Murder in Amsterdam: The Death of Theo van Gogh and the Limits of Tolerance* (New York: Penguin, 2006). The argument runs throughout the book's central analytical sections; see especially Buruma's treatment of the parallel between Ayaan Hirsi Ali and Mohammed Bouyeri as "two different visions of the universal, one radically secular, the other radically religious."

[216] Robert O. Paxton, *The Anatomy of Fascism* (New York: Vintage, 2005), 82, 91–92, 104–07.

[217] Benjamin Lee and Kim Knott, "More Grist to the Mill? Reciprocal Radicalisation and Reactions to Terrorism in the Far-Right Digital Milieu," *Perspectives on Terrorism* 14, no. 3 (June 2020): 98–115, https://www.jstor.org/stable/26918303.

Movements seeking to consolidate power or unify their followers may create or amplify "imagined enemies." These perceived adversaries serve as rallying points, galvanizing supporters by providing a clear, often menacing, target for collective anxiety and action.

A stark example of this phenomenon can be seen in Nazi Germany, where the regime propagated the narrative of an international Jewish conspiracy controlling global finance and politics. This antisemitic narrative was instrumental in uniting disparate groups under the Nazi banner and justifying horrific policies.

In contemporary society, similar dynamics can be observed, though the labels and targets have shifted. Political and ideological opponents are frequently branded with sweeping terms such as "Fascist," "White Christian Nationalist," "Zionist," "Radical Islamist," "Globalist," "Anarchist," or "Communist." The imagined enemy may be entirely fictional, grossly exaggerated, or applied indiscriminately.

This over-generalization transforms complex individuals and movements into monolithic threats, making it easier to mobilize opposition and justify extreme measures. This tactic not only simplifies the narrative for supporters but also stifles nuanced discussion and undermines the possibility of genuine understanding or compromise. Ultimately, the creation of such enemies serves more to strengthen in-group cohesion than to address real societal challenges, often at the cost of truth and social harmony.

During the first Trump administration it was not uncommon to hear allusions to Germany in the 1930's, warning that the administration was somehow equivalent to the Nazis and that President Trump was attempting to impose a fascist dictatorship. Others, from the "Right," claimed the U.S. was like Weimar Germany due to the weakness of democracy and because of the liberal and sexually permissive social climate. While rejecting both perspectives there are in fact comparisons to be made. Radical leftists engaged in street battles with radical rightists, and liberal democracy is struggling to contain the madness.

Political Violence

It is the street battles between two radical revolutionary movements and between nationalists and internationalists to which the most relevant comparisons to the Weimar era are to be found. The resurgence of White Power or the alt right was facilitated by and in reaction to the radical left. Organizations such as the Proud Boys have their *raison d'etre* in confronting the leftist Antifa organizations. The very name Antifa, anti-fascist, states the mission of the left is to oppose the radical right. Both the far left and the alt right can trace their development from the late 60's early 70's and indirectly to the street battles of Weimar Germany.

The European fascists of the early 20th century grew out of and in reaction to communists and socialists in Italy and Germany. The Communists in Germany, led by among others, Rosa Luxemburg, broke from the more moderate Social Democrats, whom they considered to have betrayed the socialist/communists international. Luxemburg was instrumental in forming the Spartacus league which would later become the Communists party of Germany.

SPARTACUS VERSUS SIEGFRIED: THE GERMAN REVOLUTION OF 1918

Siegfried, the tragic hero of Richard Wagner's *Ring Cycle*, is a fitting symbol for the National Socialist vision of Germany.[218] The final opera of the cycle, *Götterdämmerung* (*Twilight of the Gods*), serves to reflect the catastrophic downfall of Hitler's rule. Just as Siegfried catalyzed the destruction of the gods in Wagner's mythological world, Hitler precipitated the ruin of his nation and people.

In contrast, Spartacus, a Roman gladiator of probable Thracian origin—led a historic slave revolt against the Roman Republic. His legacy was revived by the revolutionary communist Rosa Luxemburg, whose organization was named the Spartacus League. Spartacus came to represent the internationalist struggle against the capitalist system that, in the eyes of the revolutionaries, had enslaved the German working class.

Germany's defeat in World War I set the stage for a period of revolution. European socialists had hoped that workers across the continent would reject the war and unite in solidarity. However, this hope proved naive, as the moderate German Social Democrats chose to support the war effort. Their decision led to a schism with radical socialists led by Luxemburg, who went on

---

[218] Alex Ross, *Wagnerism: Art and Politics in the Shadow of Music* (New York: Farrar, Straus and Giroux, 2020); David B. Dennis, "Their Meister's Voice: Nazi Reception of Richard Wagner and His Works in the *Völkischer Beobachter*," in *Opera in a Multicultural World: Coloniality, Culture, Performance*, ed. Mary I. Ingraham, Joseph K. So, and Roy Moodley (New York: Routledge, 2016), 138–50, esp. 148–49; Peter Viereck, *Metapolitics: From Wagner and the German Romantics to Hitler*, 2nd ed. (New Brunswick, NJ: Transaction Publishers, 2006); George L. Mosse, *The Crisis of German Ideology: Intellectual Origins of the Third Reich* (New York: Schocken Books, 1981); Oswald Spengler, *The Decline of the West*, trans. Charles Francis Atkinson (New York: Oxford University Press, 1991).

to form the German Communist Party (KPD), aligning closely with the Bolsheviks and the Communist International (COMINTERN).

As Germany neared defeat in the Great War, Admiral Reinhard Scheer ordered a desperate final naval attack:

> *An honorable battle by the fleet—even if it should be a fight to the death—will sow the seed of a new German fleet of the future. There can be no future for a fleet fettered by a dishonorable peace.* [219]

But the sailors mutinied. The revolt began at the naval base in Kiel and quickly spread to major cities as industrial workers joined the uprising. The German government collapsed, Kaiser Wilhelm II abdicated, and the Weimar Republic was born.[220] In reality, the Kaiser's abdication was engineered by General Paul von Hindenburg, who would later become Chancellor of the Weimar Republic. Following elections, the moderate Social Democrats gained a parliamentary majority and took responsibility for signing the Treaty of Versailles, making them the scapegoats for Germany's humiliating defeat, even though it was the Kaiser's military leadership that had led the nation into disaster. Hindenburg further encouraged the false narrative that Germany's military had been "stabbed in the back" by politicians.

During the war, the Spartacists had worked to undermine the war effort through strikes and protests, aligning themselves with the Independent Social Democratic Party.[221] When the Social Democrats joined the new government under Hindenburg, the Spartacists broke away and, on December 31, 1918, formally established the Communist Party of Germany (KPD).[222] Inspired by the Russian Revolution, they attempted to stage a similar uprising.

Rosa Luxemburg's revolutionary rhetoric, echoing the urgency of class

---

[219] History.com Editors, "German Sailors Begin to Mutiny," *History* (A&E Television Networks), updated October 27, 2020, originally published November 5, 2009, accessed January 23, 2021, https://www.history.com/this-day-in-history/german-sailors-begin-to-mutiny.

[220] The last of the Caesars; Kaiser Wilhelm II and Czar Nicholas II. Nichols abdicated on March 15, 1917, and was later executed. Wilhelm abdicated on November 11th the day the war ended. Fearing execution, he fled to Belgium where he remained in exile until his death.

[221] Marxists Internet Archive, "Spartacus League," in *Encyclopedia of Marxism: Glossary of Organizations*, accessed November 1, 2025, https://www.marxists.org/glossary/orgs/s/p.htm#spartacus-league.

[222] Rosa Luxemburg, "On the Spartacus Programme," December 31, 1918, trans. Eden and Cedar Paul, Marxists Internet Archive, accessed November 1, 2025, https://www.marxists.org/archive/luxemburg/1918/12/30.htm#doc-1.

struggle, could easily be heard at a modern-day Antifa rally:

Law! Order! Order! Law! This is the cry resounding from all sides... A strident outcry against the bogey of 'anarchy'...the well-known infernal music of a bourgeoisie concerned for its fireproof safes, its property and its profits....[223]

Yet the Social Democrat-led government, now allied with the Army, enlisted nationalist paramilitary groups known as the Freikorps to crush the uprising. More than a thousand communists died in street fighting, and many others were executed in the subsequent "White Terror."[224] Luxemburg herself was murdered, along with other leaders of the KPD.[225]

While Luxemburg tried to imitate Lenin's revolution, Hitler aimed to replicate Mussolini's. The Italian fascist leader had seized power in 1922 with his "March on Rome." Hitler's attempt, known as the Beer Hall Putsch, was swiftly quashed in Munich in 1923.[226]

After the failed putsch, communists abandoned the Spartacist legacy and rebranded themselves as anti-fascists. Their language, imagery, and ideological framework would later be revived by the American-based Antifa movement in the late 20th and early 21st centuries. Meanwhile, the Nazis shifted tactics, focusing on grassroots organizing, electoral strategy, and alliances with German conservatives. Both the Nazis and Communists gained parliamentary representation and, at times, cooperated to undermine the Social Democrats, while violently clashing in the streets.

During his imprisonment for the Beer Hall Putsch, Hitler wrote *Mein Kampf*, in which he blamed Jews, international capitalism, and communism for Germany's suffering.[227] His movement was spearheaded by the Brownshirts (SA), who used intimidation and violence to silence opponents. Escalating political instability and economic turmoil led to Hitler's appointment as

---

223 Rosa Luxemburg, "The Beginning," *Die Rote Fahne*, November 18, 1918; reprinted in *Rosa Luxemburg: Selected Political Writings*, ed. Robert Looker (London: Jonathan Cape, 1972), 253–57.

224 Rob Sewell, *Germany: From Revolution to Counter-Revolution* (London: Wellred Books, 1988), available at Marxists Internet Archive, accessed November 1, 2025, https://www.marxists.org/subject/germany-1918-23/sewell/index.htm.

225 Internationalist Communist Tendency, "A Hundred Years Since the Murder of Luxemburg and Liebknecht," *libcom.org*, January 18, 2019, accessed November 1, 2025, http://libcom.org/blog/hundred-years-murder-luxemburg-liebknecht-18012019.

226 Richard J. Evans, *The Coming of the Third Reich* (New York: Penguin Publishing Group, 2003), 192, Kindle.

227 Ibid, 24.

Chancellor. He was granted emergency powers under Article 48 of the German Constitution, powers that were ripe for abuse:

> *Laws should be made with the bad man in mind, not with the good man in mind. A good man might not abuse emergency power, but a bad man has no compunction.*[228]

Although the Nazis called themselves socialists, their ideology was nationalist, not internationalist like the communists. Right-wing paramilitaries such as the Stahlhelm (Steel Helmets), were drawn to Nazism due to their fierce opposition to communism. As Hitler consolidated power, a fire broke out in the Reichstag (parliament building), which the Nazis blamed on the communists, triggering widespread persecution.

Communists weren't the only targets. In the *Night of the Long Knives*, Hitler eliminated key rivals within the Nazi Party, cementing his absolute control. The Nazi regime ruled through terror, deporting political opponents and marginalized groups to concentration camps, eventually transforming them into death camps. Hitler's goal was to construct a racially "pure" Aryan society, the so-called Third Reich.[229]

While fascism and Nazism share many features, authoritarianism, militarism, and nationalism—only Nazism incorporated a racial ideology centered on Aryan supremacy. Nationalists were not always Nazis or fascists, but fascists and Nazis were always nationalists. Fascist ideology spread beyond Italy and Germany, influencing authoritarian regimes in Spain, Hungary, and Croatia, and even inspiring sympathizers in Britain and the United States. Both Nazi Germany and Fascist Italy pursued imperial expansion with devastating consequences.

After World War II, Nazism was widely discredited and pushed to society's fringe.

Communism, on the other hand, persisted as the principal ideological rival to liberal democracy, and, paradoxically, has seen a resurgence in Western discourse since the fall of the Soviet Union.

RTVC Table: Spartacus League vs. Nazi Party

---

[228] Benjamin Carter Hett, *The Death of Democracy: Hitler's Rise to Power and the Downfall of the Weimar Republic* (New York: Henry Holt and Company, 2018), 47.

[229] Reich, or Empire, the first was the Holy Roman Empire the second was the German/Prussian Empire.

RTVC Stage	Spartacus League / KPD	NSDAP / Nazi Party
Ideological Development	✓☐ Marxist revolutionary ideology led by Rosa Luxemburg. Aimed to dismantle capitalist structures and overthrow the Weimar regime.	✓☐ Nationalist–racial ideology grounded in antisemitism, anti-Marxism, and authoritarianism (*Mein Kampf*).
Vanguard	✓☐ Small militant revolutionary cells. Spartacus League evolved into the KPD and used strikes and agitation during WWI and 1918–1919.	✓☐ Brownshirts (SA) acted as the Nazi militant vanguard. Used intimidation, propaganda, and street violence to grow influence.
Mass Civil Unrest / Revolution	✓☐ Participated in the 1918–1919 revolution and general strikes. Attempted a Bolshevik-style uprising.	✓☐ Exploited post-war economic crises and social unrest to gain public support. Clashed with communists in the streets.
Revolutionary Victory	✗ Failed to seize control; brutally suppressed by Freikorps during Spartacist Uprising.	✓☐ Legally appointed to power in 1933 amid national crisis. Used institutional pathways to gain authority.
Tyranny / Consolidation	✗ Movement dismantled; leaders assassinated (e.g., Luxemburg).	✓☐ Purges (*Night of the Long Knives*), expanded secret police (Gestapo), propaganda, and concentration camps.
Export of Terror / Expansion	✓☐ Connected to Soviet COMINTERN, aiming for global proletarian revolution.	⚠☐ Indirect export of ideology via fascist alliances (e.g., Spain, Hungary, England). No structured global terror apparatus like COMINTERN. Military conquest.
Civil War (Optional	✗ No civil war; experienced urban revolts but not full	✗ Street violence occurred, but no organized armed

RTVC Stage	Spartacus League / KPD	NSDAP / Nazi Party
Branch)	territorial control or military parity.	resistance that formed a civil war.
Coup d'État (Optional Branch)	⚠️ Kaiser's abdication and SPD takeover can be viewed as a soft coup; lacked popular revolutionaries in leadership.	⚠️ Hitler's rise through legal means masked elite manipulation and backroom deals. Secured emergency powers under Article 48.
Democratic Transition (Foreclosed)	✗ No peaceful reform off-ramp for KPD. They were excluded and violently suppressed.	✗ Destroyed the Weimar Republic from within; ended democratic governance entirely.

WEATHER UNDERGROUND TO ANTIFA: THE AMERICAN FAR LEFT

A single spark can start a prairie fire.
— Weather Underground Manifesto

American Progressive radicals have long been divided between two opposing ideologies: communist authoritarianism and radical democracy, including anarchism.[230] Stalin's brutal dictatorship shook leftist idealism, especially among those who had once celebrated the Russian Revolution as a triumph for the working class. Stalin's non-aggression pact with Nazi Germany further disillusioned American leftists and tarnished the Communist Party's legitimacy in the U.S. Although the USSR's later alliance with the U.S. during World War II partially rehabilitated the movement, it never regained its earlier strength.

For those raised during the Cold War, the reemergence of Marxism in 21st-century America may seem baffling. The fall of the Soviet Union exposed communism as a failed social experiment. After decades of marginalization—what some have called the "fringe of the fringe"—far-left ideologies have regained traction.[231] Even after revelations that CPUSA received $40 million

[230] Howard Brick and Christopher Phelps, *Radicals in America: The U.S. Left Since the Second World War* (New York: Cambridge University Press, 2015), 260.

[231] Robert A. Fein and Bryan Vossekuil, *Protective Intelligence and Threat Assessment Investigations: A Guide for State and Local Law Enforcement Officials* (Washington, DC:

from the USSR between 1970 and 1991, and despite the imprisonment of communist-backed terrorists in the 1970s and '80s, figures like Bernie Sanders—an open socialist—have risen to national prominence.[232] Anarchists and Marxist-aligned radicals are once again finding mainstream platforms.

Several factors may explain this revival. A younger generation, unfamiliar with the Cold War's ideological stakes, may be more receptive. Academia, long a fertile ground for radical thought, continues to propagate leftist ideology. Additionally, Cold War victors may have grown complacent. Ironically, many on the Left interpret the fall of the Iron Curtain not as proof of communism's failure, but as a victory for grassroots revolution, embodied by movements like Poland's Solidarity, a labor uprising that helped bring down a communist regime.[233]

The Movement of Movements

Contemporary leftist activism is not a monolith. As Naomi Klein describes, it's a "movement of movements," a loose coalition of identity groups, environmentalists, labor activists, and anti-capitalists.[234] Rather than forming a centralized party, today's far left builds networks and coalitions.

These coalitions amplify a range of causes: racial justice, economic equality, climate activism, and opposition to various perceived forms of oppression. Central to this is a strong "anti" posture, anti-fascism, anti-capitalism, and more recently, anti-Trumpism. As in the 1960s, modern radicals exploit periods of social unrest to advance their goals, building momentum off existing progressive movements such as civil rights and anti-war activism. Veterans of past militant organizations maintain influence, linking the revolutionary fervor of the 1970s to present-day radicalism.

While a full revolutionary takeover has not occurred in the U.S., the radial strategy appears to follow familiar stages. Given that this movement is ongoing, its phases often overlap, making it a compelling case for Revolutionary Terrorism Value Chain (RTVC) analysis.

IDEOLOGICAL DEVELOPMENT

U.S. Department of Justice, Office of Justice Programs, National Institute of Justice, 1998), NCJ 170612, quoted in Randy Borum and Chuck Tilby, "Anarchist Direct Actions: A Challenge for Law Enforcement," *Studies in Conflict & Terrorism* 28, no. 3 (May 2005): 201–23, https://doi.org/10.1080/10576100590928106.

[232] Brick, *Radicals in America,* 268-269.

[233] Ibid, 266.

[234] Ibid, 289.

This connection between ideology on one hand and strategy and tactics on the other is one of the hallmarks of transformative organizing. Ideology is a worldview that gives the most essential direction to your life and leads organizations in their work to change conditions in the world.[235]

American radicalism borrows heavily from 19th- and 20th-century European Marxists, as well as Mao Tse-Tung and Latin American revolutionaries. Carlos Marighela's urban guerrilla tactics, Saul Alinsky's grassroots organizing and social disobedience tactics, and Che Guevara's *foco* theory (as expanded by Régis Debray) have profoundly influenced U.S. insurgent ideology.[236] Debray proposed that a small, disciplined vanguard could ignite revolution through symbolic, violent acts, "propaganda of the deed."

This foundational ideology manifests in at least three prominent threads in the American context:

Black Revolutionary Lineage

Figures like W.E.B. Du Bois and Marcus Garvey laid early ideological groundwork, often in dialogue with Marxist ideas. Malcolm X mocked nonviolent protest as naive, while George Jackson's *Blood in My Eye* advocated violent revolution in defense of Black liberation.[237] More recently, groups like Black Lives Matter carried forward this legacy. Some members have openly justified violence against police as a revolutionary tactic.[238]

White Intellectual Radicalism

This group includes university-educated revolutionaries who used academic and theoretical platforms to legitimize political violence. In 1965, *A Critique of*

---

[235] Eric Mann, *Playbook for Progressives: 16 Qualities of the Successful Organizer* (Boston: Beacon Press, 2011), xi.

[236] Che's Foco is adapted from Lenin's Vanguard theory. Lenin advocated the need for professional revolutionaries to lead the revolution as opposed to a more grass roots worker led uprising.

[237] It maybe that Malcolm turned away from violence, but he was assassinated before we can be sure.

[238] Jim Hoft, "Caught on Video⇒ Black Lives Matter Leader Calls for Running Over, Shooting Police," *The Gateway Pundit*, July 12, 2016 (updated August 1, 2018), accessed November 1, 2025, https://www.thegatewaypundit.com/2016/07/caught-video-black-lives-matter-leader-calls-running-shooting-police/.; Ian Tuttle, "More Hypocrisy from Black Lives Matter," *National Review*, July 12, 2016, accessed November 1, 2025, https://www.nationalreview.com/2016/07/black-lives-matter-hypocrisy-cheering-violence/.

Pure Tolerance argued that intolerance and violent action toward anti-socialists was not only justified but necessary.[239] As the fervor of the 1960s waned, Bill Ayers and the Weather Underground attempted to reignite the revolution with their 1974 manifesto *Prairie Fire*, which proclaimed:

The only path to the final defeat of imperialism and the building of socialism is revolutionary war...It includes mass struggle and clandestine struggle, peaceful and violent, political and economic, cultural and military, where all forms are developed in harmony with the armed struggle.[240]

Despite the years that have passed, Ayers has not abandoned the radical views of the Manifesto. In 2016 he appeared with his wife and fellow Weather veteran, Bernadine Dorne, on stage at The Sanctuary for Independent Media in Troy, NY. Ayers was promoting his book, *Demand the Impossible! A Radical Manifesto*, and read an excerpt;

> *What if? That humble question might be the single spark*[241] *that can ignite a massive* ***prairie fire****, provoking us to leap beyond personal speculation and into the vortex of political struggle and social action.*[242]

At the 2016 event Dorne spoke of supporting the Black Lives Matter organization and the occupy movement. She stated these movements were not spontaneous but organized over time and stressed the importance of connecting key issues. Weather veterans have been instrumental in creating this movement of movements uniting diverse organizations toward the same goal.

Eric Mann, while never going underground, was a founding member of the

---

239 Herbert Marcuse, "Repressive Tolerance," in Robert Paul Wolff, Barrington Moore Jr., and Herbert Marcuse, *A Critique of Pure Tolerance* (Boston: Beacon Press, 1969), 109–10.

240 Weather Underground, *Prairie Fire: The Politics of Revolutionary Anti-Imperialism: Political Statement of the Weather Underground* (San Francisco: Communications Co., 1974), 2, PDF, https://www.sds-1960s.org/PrairieFire-reprint.pdf. sds-1960s.org.; Brian Burrough, *Days of Rage: America's Radical Underground, the FBI, and the Forgotten Age of Revolutionary Violence* (New York: Penguin Press, 2015), 312.

241 Spark, or Iskra, was the title of a journal edited by Lenin in 1900s, presumably an inspiration for Ayers. Lenin's first newspaper had been called *Iskra*, or "spark," after a line one poet, Alexander Odoevsky, wrote in response to another, Pushkin: "from a spark a fire will flare up."; Ted Widmer, "Lenin and the Russian Spark," *The New Yorker*, April 20, 2017, https://www.newyorker.com/culture/culture-desk/lenin-and-the-russian-spark.

242 Bill Ayers and Bernardine Dohrn, "Bill Ayers and Bernardine Dohrn 'Demand the Impossible,'" YouTube video, posted by *media sanctuary*, accessed November 1, 2025, https://www.youtube.com/watch?v=ToBX77nTXyI, at 4:49.

Weather organization. His 2011 book, *Playbook for Progressives*, is an obvious reference on Saul Alinsky's *Rules for Radical.*[243] Mann's book is described as "An Art of War for organizers."[244] In it he describes the importance of creating coalitions.[245] Mann was instrumental in helping found the Black Lives Matter organization and is the inspiration behind the accusation that the BLM founders "are trained Marxists." In a December 2020 video, BLM founder Patrisse Cullors states, "I do believe in Marxism."[246] And in a 2018 interview she stated she was a trained by Eric Mann.

> PATRISSE KHAN-CULLORS: *Well, I'm a trained organizer. And so, I think sometimes people think that because Black Lives Matter is the biggest thing, that that's the first thing I ever did. And it's not. I was trained knocking on doors, you know, getting on buses and passing out flyers and getting people to join organizations. The Labor Community Strategy Center is my first political home. It's where I would be a part of what it's famous for, which is the Bus Riders Union.*
>
> JUAN GONZÁLEZ: *Started by an old friend of mine, Eric Mann.*
>
> PATRISSE KHAN-CULLORS: *Yes, Eric Mann. That's my mentor.*

Eric Mann, director of the Labor Community Strategy Center, had been a member of Students for a Democratic Society (SDS) in the 1960's.[247] When SDS dissolved he joined the weathermen. But when select members of the organization went underground Mann did not, "I remember Bill Ayers said to me, "Eric, you're too big and too visible to go underground."[248]

Mann had been involved in several well publicized violent events in September and November of 1969 resulting in his being charged with assault

---

243 Mann, *Playbook for Progressives.*

244 Mike Davis, endorsement quoted in "Playbook for Progressives."

245 Mann, xx, 198.

246 Patrisse Cullors, "Am I A Marxist?," *YouTube*, December 14, 2020, video, accessed November 1, 2025, https://www.youtube.com/watch?v=rEp1kxg58kE. UTUPub+1; Patrisse Khan-Cullors, interview by Amy Goodman and Juan González, "'When They Call You a Terrorist': The Life of Black Lives Matter Co-Founder Patrisse Khan-Cullors," *Democracy Now!*, January 16, 2018, transcript, accessed November 1, 2025, https://www.democracynow.org/2018/1/16/when_they_call_you_a_terrorist.

247 Labor/Community Strategy Center, "People," *The Labor Community Strategy Center*, accessed November 1, 2025, https://thestrategycenter.org/about/people/; and "About," accessed November 1, 2025, https://thestrategycenter.org/about/.

248 Jonah Raskin, "Eric Mann: Public Organizer," *Counterpunch*, (June 2018): accessed November 14, 2025, https://www.counterpunch.org/2018/06/01/eric-mann-public-organizer/.

and battery, and disturbing the peace from a 25 September disturbance and in November during a protest by the Weathermen,

> *The demonstrators, carrying a large NLF flag, stood on the front steps of the building chanting "Power to the people" and "Off the pigs," then attempted to enter the building. Three Harvard police told the group that the building was locked, and refused to let them in.*[249]

At the time of the November demonstration Mann was out on bail.

> CAMBRIDGE, Mass., Nov. 22 -- *Eric Mann, the leader of the militant Weatherman faction of the Students for a Democratic Society, was free today in $4,000 bail after his arrest on charges of conspiracy to commit murder, assault with intent to commit murder, threatening and the promotion of anarchy.*[250]

Anti-imperialism was integral Weather's ideology and Mann hadn't given up on the rhetoric. In 2006 6e wrote that U.S. imperialism has deteriorated "into its most militaristic and decadent form."[251] Fellow Weather veteran Michael Novick shares Mann's continued espousal of communist anti-imperialist rhetoric.

> *Civil war is not a future prospect, but a current reality, constituting a form of blow-back or introgression into the settler colonial metropole of the US empire of the strategy and tactics of low-intensity warfare that the US has been using for decades to undermine anti-colonial and anti-capitalist uprisings… The left needs to engage with a firmly anti-capitalist and anti-imperialist politics…* [252]

Michael Novick has been a major player in organizing Antifa. His organization Anti Racist Action LA formed in 1987 to "fight fascism,

---

[249] Shirley E. Wolman, "SDS and Weathermen Hold Separate Protests," *The Crimson*, (November 1969): accessed November 14, 2025, https://www.thecrimson.com/article/1969/11/26/sds-and-weathermen-hold-separate-protests/ ; https://www.thecrimson.com/article/1969/10/30/mann-weathermen-released-after-arrests-for/

[250] "Weathermen in SDS Accused of Attack on Cambridge Police," *New York Times*, November 23, 1969, https://www.nytimes.com/1969/11/23/archives/weathermen-in-sds-accused-of-attack-on-cambridge-police.html.

[251] Berger, Dan, *Outlaws of America: The Weather Underground and the Politics of Solidarity*, (Oakland: AK press, 2006), 81; Mann, *Playbook for Progressives*, xxi.

[252] Michael Novick, "Low Intensity Civil Warfare is Underway in the US," *Turning The Tide*, 34, no. 1 (Jan-March 2022):accessed January 21, 2022, Low Intensity Civil Warfare is Underway in the US – Fighting fascism, colonialism, and white supremacy (antiracist.org)

colonialism, and white supremacy."[253] This organization formed the torch network as a hub coordinating the activities of the greater antifa community.[254]

Weather's legacy has also impacted academia, Ayers became Professor of education at University of Illinois at Chicago, and Dorne taught at Northwestern University school of law; Susan Rosenberg taught literature at the City University of New York. Other Academics shared the ideological drive and used their position to justify violence in the name of revolutionary change. Ward Churchill was notable for his call to violent revolution in his 1998 essay, *Pacifism as Pathology.* Churchill not only condoned violent revolution but insisted non-violent resistance is pathological.[255]

Ayers has never renounced their years as terrorist nor the goal of revolution. In recent years, Ayers has focused on mentoring younger activists and educators and engaging in public discussions on the role of education in democracy and resistance. He speaks regularly at colleges, protests, and academic panels, especially in education programs with activist leanings. Ayers has expressed support for student-led movements, including Black Lives Matter, climate strikes, and campus labor struggles.[256]

The White Working Class

The third and arguably least influential line of radicals is the white working class. In the '60s and 70,' a group named the Young Patriots emerged in Chicago and worked closely with the Black Panther Party.[257] Following in their tradition a small group calling itself Redneck Revolt demands revolution and promotes armed resistance.[258] Although Redneck Revolt appears almost farcical, their rhetoric is nevertheless worrisome. From the Redneck Revolt website, note that

---

253 Berger, Dan, *Outlaws of America*, 314.; Anti-Racist Action–Los Angeles/People Against Racist Terror (ARA-LA/PART), "Fighting fascism, colonialism, and white supremacy," accessed November 20, 2025, https://antiracist.org/.

254 Ngo, Andy, *Unmasked: Inside Antifa's Radical Plan to Destroy Democracy* (New York: Hatchette Book Group, 2021), 83.

255 Brick, *Radicals in America*, 288.

256 Bill Ayers, interview by Aditi Patwardhan, "Action and Activism with Bill Ayers," *The Science Survey*, February 12, 2025, https://thesciencesurvey.com/spotlight/2025/02/12/action-and-activism-with-bill-ayers/.

257 Patrick King and William "Preacherman" Fesperman, "Young Patriots at the United Front Against Fascism Conference (1969)," *Viewpoint Magazine*, August 10, 2015, https://www.viewpointmag.com/2015/08/10/young-patriots-at-the-united-front-against-fascism-conference/.

258 Redneck Revolt, "Principles," *Redneck Revolt*, accessed November 20, 2025, https://www.redneckrevolt.org/principles.

violence is not to violate the revolutionaries' basic humanity, no such guarantee for those they intend to turn their guns on. It is doubtful they see the hypocrisy of their position.

It is time to turn our guns on our real enemies.

WE BELIEVE IN THE RIGHT OF MILITANT RESISTANCE

We are not pacifists. We believe in using any and all means at our disposal to gain our freedom and true liberty, provided those methods do not violate our basic humanity or integrity.

WE BELIEVE IN THE NEED FOR REVOLUTION

We believe that there will have to be a complete restructuring of society to provide for the survival and liberty of all people.

Climate and Environmental Activists

Environmental activists may be a fourth line of leftist radicals though it can be difficult to separate those who have joined the movement of movements. The 2004 book *The Logic of Political Violence* by Craig Rosebraugh, spokesman for both the Animal Liberation Front and the Earth Liberation Front, laments the ineffectiveness of non-violence. He describes policeman as equivalent to Nazi storm troopers. The periodical *The Jacobin* highlights the Millennial Generation's interest in radical leftism.[259] That the publishers would name themselves after the radicals who ruled during the Reign of Terror of the French Revolution should be warning enough.

The catalyzing events propelling the Left to violence in the '60s were the assassinations of Dr. Martin Luther King and Robert Kennedy. The Left claimed these were the work of the fascist American government. Some radicals felt the non-violent efforts of the anti-war and civil rights movements had proven ineffective, and so they turned to violence. Labor and civil rights grievances in the United States in the early 20th century were real, and the country was slow to address the issues until pushed to do so by the Left.

The catalysts for violence in the 21st century was first, the 2003 invasion of Iraq, and then the election of President Donald Trump in 2016 and again in 2024. A general dislike of President Trump, support for illegal immigration and for the free Palestine movement are the unifying elements with continuing focus on racial discrimination and anti-imperialism, and environmental issue rounding out the common causes on the American Left. The stated grievances have changed with the times, and the tactics have evolved, but the goals,

---

[259] Brick, *Radicals in America*, 309-310.

overthrowing the existing order, remain the same.

Ideological cohesion is forged through revolutionary literature, manifestos, and narratives portraying the U.S. government as irredeemably fascist and white supremacist.

THE VANGUARD

One of the paradoxes of freedom is that you never feel more free than when you're fighting against an obstacle to humanity.[260] *Bill Ayers*

While many veterans of the Weather Underground have taken divergent paths, their views reflect a spectrum ranging from remorse to unapologetic militancy. In a recent documentary, *Mark Rudd*, once the student leader at Columbia University, expressed regret for his involvement. By contrast, *Bill Ayers*, now a professor of education, remains defiant. In a 2001 *New York Times* profile, ironically published on September 11th, Ayers stated: *I don't regret setting bombs… In fact, I don't think we did enough.*[261] This frank admission undercuts claims that the Weather Underground sought to avoid harming people. *Howie Machtinger*, a fellow radical, confirmed: *The myth is that Weather never set out to kill people… and it's not true. Policemen were fair game.*[262]

The Weather Underground's lineage can be traced to figures like *Sam Melville*, who carried out a bombing campaign in New York City in 1969, hoping to emulate the anarchists of the twentieth century. Anarchists had bombed Wall Street in 1920, killing over thirty people. Although quickly arrested, Melville became a martyr following his death during the Attica prison uprising in 1971—his legacy inspired the Weather Underground Organization (WUO).

WUO evolved from the *Students for a Democratic Society (SDS)* in 1969. They saw themselves as the *foco*, a revolutionary vanguard dedicated to jump-starting a communist revolution in the U.S. through armed struggle. Adopting urban guerrilla tactics from *Carlos Marighela*, WUO carried out bombings at the U.S. Capitol and Pentagon. Yet, the broader revolution never materialized. While some were captured, many simply surrendered as it became clear that public

---

[260] Bill Ayers, interview by Aditi Patwardhan, "Action and Activism with Bill Ayers," *The Science Survey*, February 12, 2025, https://thesciencesurvey.com/spotlight/2025/02/12/action-and-activism-with-bill-ayers/.

[261] James Miller, "Return of the Weathermen: The unhappy afterlife of '60s radicalism," The Boston Globe 10/19/2003 http://archive.boston.com/news/globe/ideas/articles/2003/10/19/return_of_the_weathermen/.

[262] Burrough, *Days of Rage*, 123.

support was lacking.

Bobby Seale and Huey Newton founded the Black Panther Party for Self-Defense (BPP) in 1966, a response to deep-seated grievances in Black communities.[263] The BPP's rapid growth was fueled by its direct engagement with issues of police brutality, poverty, and systemic injustice. However, the party's momentum was short-lived; it was undermined by the FBI's Counterintelligence Program (COINTELPRO), which targeted the Panthers with infiltration and disruption. Internal divisions also emerged, with some members advocating reform while others, more radical, pushed for revolution. The most militant faction, led by figures like Eldridge Cleaver, broke away to form the Black Liberation Army (BLA), which was implicated in several police killings.[264]

As Peniel Joseph noted in *Waiting 'til the Midnight Hour: A Narrative History of Black Power in America,* "If you talk about a revolution against the state, the state responds accordingly!" [265] Most members of the BLA were eventually killed or captured.

Remnants of the Weather Underground and the BLA later united as the May 19th Communist Organization (M19CO), continuing their campaign with violent acts, including a notorious armored car robbery and the assassination of a New Jersey State trooper.[266] Joanne Chesimard, also known as Assata Shakur and the godmother of rapper Tupac Shakur, was convicted in connection with the attack but escaped prison and fled to Cuba, where she remains in exile. Cuba exerted some degree of influence, if not outright control—over the Weather Underground and other revolutionary groups, including the Puerto Rican FALN.[267]

---

[263] Brick, *Radicals in America*, 92.; Burrough, *Days of Rage*, 42.

[264] ICSR Team, "The Black Liberation Army and Homegrown Terrorism in 1970s America," *International Centre for the Study of Radicalisation*, April 12, 2012, http://icsr.info/2012/04/the-black-liberation-army-and-homegrown-terrorism-in-1970s-america/.

[265] Sean Elder, "Does Black Lives Matter Pick Up Where The Black Panthers Left Off?" *Newsweek*, October 2016: accessed October 21, 2017, http://www.newsweek.com/2016/10/14/black-panthers-black-lives-matter-505542.html.

[266] "May 19th Communist Organization (M19CO)," *GlobalSecurity.org*, last modified September 24, 2017, https://www.globalsecurity.org/military/world/para/m19co.htm.

[267] "Weather Underground (Weatherman/The Weathermen)," *InfluenceWatch*, accessed November 20, 2025, https://www.influencewatch.org/organization/weather-underground-weatherman-weathermen/.; Zach Dorfman, "How Fidel Castro

Law Enforcement Response and Legal Tactics

Law enforcement, particularly the FBI, responded with aggressive and often illegal tactics. COINTELPRO's methods were so controversial that the Socialist Workers Party successfully sued the government, exposing the extent of these covert operations.[268] Even members of the FBI's own Squad 47 faced prosecution for their actions. Judges at the time considered illegal government conduct a greater threat to society than the terrorist acts themselves.[269] The debate over the legality and ethics of government counterterrorism tactics remains relevant in contemporary struggles with extremism.

Prosecution Challenges and the RICO Strategy

Prosecuting members of radical organizations proved difficult due to the government's irregular and sometimes unlawful investigative methods. Authorities often resorted to charging suspects with lesser, more easily proven crimes simply to disrupt their operations. As a result, many Weather Underground members served little or no jail time. [270] A turning point came when the government began using the Racketeer Influenced and Corrupt Organizations (RICO) Act—originally designed to combat organized crime, against these groups. RICO prosecutions led to convictions of M19CO members and were later applied to dismantle far-right organizations as well.

The Rise and Evolution of the Black Bloc and Antifa

In the 1990s, the surge of anti-globalist and anti-capitalist protests gave rise to the Black bloc, a loosely organized collective of anarchists known for their militant tactics and striking presence at events like the "Battle of Seattle" where coordinated acts of vandalism and street clashes disrupted the World Trade Organization summit. This group, characterized by their uniform black attire and coordinated actions, became synonymous with confrontational demonstrations targeting symbols of corporate and governmental power.

Over time, the Black bloc's tactics and ethos evolved, fueling the growth of the broader Antifascist (Antifa) movement. Antifa has expanded its reach and influence, drawing on a network of loosely affiliated "affinity groups," a

---

Supported Terrorism in America," *Wall Street Journal*, June 8, 2017, https://www.wsj.com/articles/how-fidel-castro-supported-terrorism-in-america-1496961035.

[268] Burrough, Days of Rage, 374.

[269] Ibid, 495-7.

[270] Brent Smith, *Terrorism in America: Pipe bombs and Pipe Dreams (Albany:* SUNY Press 1994), 104.

concept rooted in anarchist organizing from the Spanish Civil War era.[271] Michael Novick's *Anti-Racist Action LA* (founded in 1987) played a pivotal role in shaping and mobilizing this movement.

Antifa maintains that it is not a formal organization but rather a decentralized movement united by shared principles and tactics. Most of its direct actions have centered on property destruction, street clashes, and inciting riots, though the scale and targets have varied. Notable incidents involving Antifa include violent demonstrations at the 2016 presidential inauguration in Washington, D.C. and participation in the widespread anti-police protests of summer 2020. Some commentators alleged Antifa involvement as agents provocateurs during the January 6, 2021, Capitol riot, though FBI Director Christopher Wray testified that the Bureau found no evidence of organized Antifa participation in the breach. The allegation itself, however, illustrates the "imagined enemy" dynamic discussed earlier in this chapter: each side attributes provocative violence to infiltrators from the opposing camp, reinforcing in-group cohesion while deflecting accountability.[272]

MASS CIVIL UNREST, REVOLUTION, AND CIVIL WAR

The late 1960s were a crucible of radical activism. Jerry Rubin and Abbie Hoffman, drawing from the anti-war movement, founded the Youth International Party—better known as the Yippies. Their signature weapon was *Guerilla Theatre*: provocative, media-savvy stunts designed to stir mass civil disobedience. Hoffman, acutely aware of television's power, orchestrated confrontations with police at the 1968 Democratic National Convention, aiming to ensure the spectacle would dominate headlines.[273]

Despite the era's upheaval, the radical left never managed to mainstream revolution. The conclusion of the Vietnam War and the passage of landmark civil rights laws addressed many of the most urgent grievances. As Black Liberation Army veteran Sekou Odinga later reflected, "It was all a mistake. People weren't ready for armed struggle…we were way in front of the people. You can be a vanguard in a struggle, but you have to have the people behind

---

[271] Brick, *Radicals in America*, 205.; "Affinity Groups," *libcom.org*, November 13, 2003, accessed November 20, 2025, https://libcom.org/book/export/html/12106.; "History of Affinity Groups," *Organizing for Power, Organizing for Change*, December 2, 2008, accessed November 20, 2025, https://www.organizingforpower.org/history-of-affinity-groups/.

[272] *Capitol Punishment,* directed by Christopher Burgard (Bark at the Hole Productions, 2021), https://capitolpunishmentthemovie.com/.

[273] Abbie Hoffman on Yippie Tactics (video), 1968, accessed December 21, 2017, https://www.youtube.com/watch?v=2oujcg_Tifw.

you, and they weren't".[274]

Evolving Tactics: From Armed Struggle to Strategic Nonviolence

By 2011, the conversation around protest had evolved. Robert Helvey, a former U.S. Army officer familiar with Burma's opposition movement, delivered a lecture on Strategic Non-Violent Conflict. Helvey, influenced by theorist Gene Sharp, advocated for applying military planning to peaceful resistance. His advice for protest marches was practical: keep crowds noisy to drown out the intimidation of armed security, and above all, ensure demonstrators remain visibly non-violent. The goal: let the media broadcast the brutality of the authorities, swaying public opinion in favor of the protesters. When asked about the anarchists who disrupted the WTO meeting in Seattle, Helvey surprised his audience: "Oh, you mean the boys in black? I love those guys."

The Black Bloc and the Internet Age of Protest

In 1998, Peoples Global Action convened in Geneva to strategize for the anti-WTO protests in Seattle. The 1999 "Battle of Seattle" marked a watershed moment: the first major protest organized via the Internet, drawing 40,000–50,000 demonstrators.[275] Groups like the Direct Action Network, trained in non-violent tactics by the Ruckus Society, joined forces with militants who adopted Black Bloc tactics, methods honed by German anarchists in the 1980s.[276] For the Black Bloc, property destruction was not violence; they rejected the very concept of property rights.[277]

This militant approach deepened the rift between moderate leftists and the Black Bloc, who "don't respect the democratic planning process, don't accept the majority's will and do whatever they want, drawing in youth who are inclined to be anti-authority".[278] Seattle signaled to the radical left that the tactics of the ultra-radicals from the 1960s and '70s still had resonance.[279]

Shifting Alliances and the Struggle for Mass Support

Yet, traditional labor organizations distanced themselves from the extremists. As in the 1960s, the far left struggled to win over the working class, which instead gravitated toward movements like the Tea Party and, later,

[274] Burrough, *Days of Rage*, 538.
[275] Brick, *Radicals in America*, 284.
[276] Brick, *Radicals in America*, 285-6.
[277] Ibid, 286.
[278] Ibid, 287.
[279] Ibid, 288.

Donald Trump. In a bid to mobilize the masses, Antifa aligned with Black Lives Matter, echoing the earlier alliance between the Weathermen and the Black Liberation Army. Police violence against Black Americans became a rallying point, as seen in the riots of summer 2020.

Both Antifa and BLM have served as transitional movements, normalizing confrontational tactics and lowering the barrier for ordinary people to challenge authority. The willingness of average citizens to oppose law enforcement remains a critical threshold for any revolutionary movement.

The Radical Left's Enduring Ambition

The radical left continues to see itself as the enlightened vanguard, convinced of its role in ushering in a communist utopia. For now, these groups remain on the fringe, unable to muster the mass support needed to overturn the system. Should they ever succeed, history offers ample warning signs of how a radical leftist regime might wield power.

REVOLUTIONARY VICTORY AND CONSOLIDATION OF POWER

What might happen if the radical white ideological left were to seize power in America? As Jerry Rubin once quipped, "The hippie streets would be the first cleaned up by 'socialist' pigs. We'd have to attend compulsory political-education classes at least five nights a week. The Left turns Communism into a church with priests defining 'the line.'"[280] Rubin's tongue-in-cheek warning underscores the paradox: those who championed liberation and rebellion could quickly become the new enforcers of ideological conformity.

Disturbingly, much of the Weather Underground's ideology has seeped into mainstream culture. As Dwight MacDonald observed, there's a crucial distinction between liberals, who see themselves as "the left opposition within the present society," and radicals, who seek to upend society altogether. Radicals push from the margins, while liberals enact reforms; together, they create a dynamic that gradually shifts the cultural and political landscape.[281]

The self-criticism rituals once practiced by the Weather Underground and other Marxist groups now echo in the modern cancel culture. Those who stray from the accepted orthodoxy are often subjected to public shaming and compelled to make humiliating, self-deprecating apologies. In some cases, refusal to comply has even led to legal consequences. For example, a Denver

---

[280] Jerry Rubin, *DO IT!: Scenarios of the Revolution* (New York: Simon and Schuster, 1970), accessed November 20, 2025, http://www.panarchy.org/rubin/doit.html.
[281] Berger, *Outlaws of America*, 96 & 210.; Brick, *Radicals in America*, 312.

cake maker was ordered to attend political reeducation training after declining to make a cake celebrating same-sex marriage.[282]

Should the hard left ever consolidate real power, these trends could be only the beginning. The mechanisms of ideological enforcement, once confined to the fringes—may well become the tools of a new orthodoxy, shaping society in ways both overt and subtle.

TYRANNY

It isn't difficult to envision what a leftist tyranny might look like. The brief experiment of the Capitol Hill Autonomous Zone (CHAZ) in Seattle offered a telling preview: once "the People" have spoken, dissenters find themselves at the mercy of mob rule, with no real recourse or due process. Forced conformity becomes the norm, with public calls to jail climate change skeptics and vaccine dissenters serving as chilling examples of how dissent could be criminalized.

Efforts to defund local police and replace them with federally mandated "tolerance training" are often framed as reforms, but they also lay the groundwork for a politically reliable security force, one that enforces ideological orthodoxy rather than impartial law. The prospect of government agencies being weaponized against citizens is no longer far-fetched. The IRS has a documented history of targeting political opponents, and the politicization of FBI leadership has raised serious concerns about the impartiality of federal law enforcement.[283]

In such a system, the machinery of government is no longer a neutral arbiter but an instrument of ideological enforcement, one that punishes dissent and rewards conformity.

EXPORT AND EXPANSION

The radical Marxist ideology that took root in the 1960s has found new life and fresh momentum in the 2020s. Concepts like "white privilege," now omnipresent in public discourse, were already being used by activists like Bill

---

[282] "Colo. baker faces 're-education' training after state court refuses to take case," *Catholic News Agency*, April 28, 2016, https://www.catholicnewsagency.com/news/33805/colo-baker-faces-re-education-training-after-state-court-refuses-to-take-case.

[283] Peter Overby, "IRS Apologizes For Aggressive Scrutiny Of Conservative Groups," *NPR Politics* (October, 2017):https://www.npr.org/2017/10/27/560308997/irs-apologizes-for-aggressive-scrutiny-of-conservative-groups?t=1646745150184.; Joseph E. diGenova, "The Politicization of the FBI," *Imprimus* 47, issue 2 (February 2018): https://imprimis.hillsdale.edu/the-politicization-of-the-fbi/.

Ayers as early as 1966.[284] Today, it's not uncommon to see protest organizers brandishing Marxist symbols and invoking the language of "resistance." Critics of this convergence, including former leftist activists like David Horowitz, argue that mainstream liberal movements have been instrumentalized by more radical elements whose ultimate objectives extend well beyond reform.

One persistent challenge for American radicals has been their inability to build the mass support necessary to overthrow the system. Fragmented into competing factions, each with its own pet grievance, the far left has struggled to unite under a single banner. To address this, they have engineered a "movement of movements," forging intersectional alliances among groups fighting for climate action, anti-racism, anti-capitalism, anti-fascism, and anti-Trumpism. The election of President Trump in 2016 provided a common adversary, galvanizing these disparate elements and helping to bring radical ideas from the margins into the mainstream.

Yet another obstacle for the radicals is the enduring contentment of the average American. Most people continue to enjoy relative freedom and prosperity, leaving them largely unmoved by calls for revolution. While leftist elites may feel alienated from the system, the broader public has not experienced the kind of catalytic crisis that would drive them to embrace radical change. However, the upheaval brought on by the Covid pandemic and a sharp economic downturn may have shifted the balance, potentially swelling the ranks of those open to radical solutions.

It's important to distinguish "the system" from the so-called "establishment." Disillusionment with establishment parties is already widespread, as evidenced by the rise of outsider candidates: Obama and Bernie Sanders on the progressive side, Trump on the conservative. Yet none of these figures have truly toppled the establishment or delivered on promises to champion the average citizen. This creates a dangerous vacuum, one that radicals on both the left and right are eager to exploit, stoking unrest by amplifying both real and imagined grievances. In this volatile environment, each side feeds off the fears and anxieties of the other, fueling a cycle of polarization and instability.

Summary: The Radical Left in America

The question of designating groups like Antifa as terrorist organizations exposes the profound tension between national security and civil liberties in modern America. While a full-blown revolutionary victory for the radical left remains speculative, the ideological legacy of groups like the Weather

---

[284] Berger, *Outlaws of America*, 96 & 210.

Underground has undeniably seeped into the mainstream, shaping the language and tactics of today's activist networks, educational institutions, and media. What was once the domain of fringe revolutionaries now manifests as a diffuse cultural force, most visible in the normalization of public shaming, compelled apologies, and institutional punishments for those who stray from progressive orthodoxy. The case of the Denver baker, ordered to attend sensitivity training for refusing to make a cake for a same-sex wedding, is just one example of how bureaucratic mandates and cultural pressure can enforce conformity in the absence of overt state power.[285]

This "soft enforcement" of ideology, through cancel culture, compelled speech, and institutional capture, lays the groundwork for broader systems of control should the radical left ever achieve true political dominance. Traditional labor unions have largely distanced themselves from the far left, and the working class has often shifted rightward in response to economic and cultural anxieties. In turn, Antifa and similar groups have forged alliances with movements like Black Lives Matter, echoing earlier coalitions between the Weather Underground and the Black Liberation Army. The mass unrest of summer 2020 showcased how racial grievances and police violence can be leveraged to mobilize the public, normalize street-level confrontation, and desensitize society to political violence.

At the core of the far left's strategy is the vanguard model: clandestine militant cells like the Weather Underground and M19CO, and now decentralized networks such as Antifa and Redneck Revolt, all believe that targeted violence or disruption can serve as a catalyst for broader uprisings. Following the logic of Debray's *foco* theory, these groups see themselves as the spark that will ignite mass revolt. While outright revolution has not materialized, the line between peaceful protest and revolutionary violence has become deliberately blurred, allowing militants to hide behind legitimate civil resistance.

Though leftist radicals have not seized the levers of state power, they have articulated how consolidation would unfold: political indoctrination, the replacement of local police with ideologically loyal security forces, and the capture of key institutions. Early signs of this hypothetical tyranny are already

---

[285] Catholic News Agency, "Colo. baker faces 're-education' training after state court refuses to take case," (April 2016): accessed 13 July 2025, Colo. baker faces 're-education' training after state court refuses to take case | Catholic News Agency.; Kaelan Deese, "Gorsuch Suggests Colorado Baker Forced to go through 'Reeducation Program,'" *Washington Examiner,* (December 12, 2022): accessed 12 July 2025, Gorsuch suggests Colorado baker forced to go through 'reeducation program' - Washington Examiner.

visible: speech control through deplatforming, compelled diversity training, and mob rule in spaces like Seattle's Capitol Hill Autonomous Zone, where formal authority was replaced by ideological tribalism.

On the global stage, the radical left has succeeded in exporting its symbols, tactics, and rhetoric. Antifa and allied groups now operate transnationally, reframing local struggles as part of a worldwide battle against fascism and capitalism. Social media and international conferences amplify these efforts, making the "movement of movements" a truly global phenomenon.

While a classic coup d'état remains unlikely in the American context, the long-term strategy of institutional infiltration, gaining control over bureaucracies, media, and security agencies, presents a subtler but no less significant threat to democratic norms. The prospect of a "quiet coup," achieved through institutional conquest rather than open revolt, is a modern expression of revolutionary ambition.

In sum, the radical left's greatest success has been in shifting the cultural and institutional landscape, normalizing confrontation with authority, and eroding the boundaries between protest and insurrection. As polarization deepens and institutions falter, the risk of escalation, whether through mass unrest or institutional subversion, remains ever-present. The challenge for American democracy is to defend civil liberties and the rule of law without succumbing to the temptations of either heavy-handed repression or ideological conformity.

RTVC Table: American Radical Left

RTVC Phase	Presence in U.S. Far-Left Movements
Ideological Development	✓ Marxist, anarchist, anti-imperialist thought
Vanguard Formation	✓ Weather Underground, M19CO, Antifa networks
Mass Civil Unrest	✓ 1960s–70s riots, 2020 protests, BLM actions
Civil War (Branch)	✗ Not yet—though anticipated by some ideologues
Revolutionary Victory	✗ Hypothetical; not yet achieved

RTVC Phase	Presence in U.S. Far-Left Movements
Coup d'État (Branch)	✚ Theorized; infiltration of institutions noted
Tyranny	⚠ Not applicable. However, the logic of vanguard consolidation suggests that revolutionary success would require coercive mechanisms inconsistent with constitutional pluralism, as demonstrated in analogous cases (Chapter 1).
Export & Expansion	✓ Transnational activism, Antifa global network
Democratic Off-Ramp	✗ Rejected or undermined by radical factions

THE ORDER TO THE PATRIOT FRONT: THE AMERICAN FAR RIGHT

The tree of liberty must be refreshed from time to time with the blood of patriots an d tyrants. — Thomas Jefferson

The Ku Klux Klan no longer holds the power it once wielded in the early 20th century. While the Klan historically claimed to uphold the founding principles of the United States, particularly states' rights, it did not extend that defense to individual rights, especially for minorities. Ironically, Klansmen fought against the Nazis during World War II, perceiving themselves, however mistakenly, as defenders of the American way of life.

In contrast the modern White Power movement is rooted in a fantasy of racial separatism and civil war. Its goal is the creation of a utopian ethno-state through violent upheaval.[286] Following the Vietnam War, figures like Louis Beam, a combat veteran, became disillusioned with the U.S. government and began spreading the narrative that American soldiers had been "sold out" by politicians, an echo of the "stab-in-the-back" myth that fueled German fascism after World War I.

[286] Kathleen Belew, *Bring the War Home: The White Power Movement and Paramilitary America.* (Cambridge: Harvard University Press, 2018), 5.

IDEOLOGICAL DEVELOPMENT

The White Power movement is overtly anti-American. It aims to dismantle what it has named the "Zionist Occupation Government (Z.O.G.)" and the "New World Order."[287] Scholar Kathleen Belew describes this shift by defining "White Power" as a distinct movement that defies the conventional left-right political spectrum. She argues, as does this author, that the radical left and radical right often share more in common with each other than with their respective mainstreams.[288] Belew identifies three core factions within the White Power revolution:

- Neo-Nazis
- Christian Identity adherents
- Anti-federalist extremists

Neo-Nazis and the Shift in Ideological Ground

In the aftermath of World War II, Nazism was not only discredited but decisively defeated. American neo-Nazis were largely marginal until the mid-1980s, when attempts to revive Aryan supremacism began gaining limited traction. Most adherents were disenfranchised older men, in sharp contrast to the affluent white youth seen in left-wing activism. Many Neo-Nazis were career criminals who blurred the line between ideological commitment and organized crime, dragging the movement into drug trafficking and prostitution.

Christian Identity and Pagan Revivalism

Imported from Britain, Christian Identity theology, also known as "replacement theology," posits that Aryans are the true chosen people. This ideology parallels other supremacist doctrines such as Islamic jihadism or Soviet visions of the "New Man." Despite its Christian foundations, the movement has seen a growing shift toward paganism, particularly Norse mythology, which appeals to white racial mysticism.[289]

Though ideologically incompatible, Christian Identity and pagan factions coexist uneasily. Both view apocalyptic conflict as necessary: Christian Identity radicals anticipate Armageddon, while pagans await Ragnarök. Both hope for global collapse as a pathway to establishing a racially pure utopian world

---

[287] Ibid, 7.

[288] Ibid, ix.

[289] Mattias Gardell, (2003-06-26T23:58:59). *Gods of the Blood: The Pagan Revival and White Separatism* (London: Duke University Press, 2003), location 160-178, Kindle.

order.[290]

The White Power movement merged with older anti-federalist beliefs, especially among tax protestors and constitutional purists. They often invoked Jeffersonian rhetoric about the "tree of liberty" needing to be "watered with the blood of patriots and tyrants."

Anti-federalist extremists

The Sovereign Citizen Movement is a loosely organized, anti-government ideology that has gained traction on the fringes of American society. Its adherents believe that they are answerable only to their own interpretation of "common law" and are not subject to federal, state, or local statutes. Rejecting the legitimacy of the U.S. government, sovereign citizens frequently employ pseudo-legal jargon and convoluted paperwork in an attempt to sidestep taxes, licenses, and even criminal charges. Their tactics range from filing bogus liens against public officials to refusing to recognize the authority of courts and law enforcement, often resulting in tense, and sometimes violent, confrontations with authorities.[291]

What makes the Sovereign Citizen Movement particularly concerning is its capacity to exploit the cracks in America's legal and political systems. While most Americans may view government bureaucracy as a necessary, if frustrating, part of civic life, sovereign citizens see it as an illegitimate force to be resisted at every turn. The movement's appeal has grown during periods of economic uncertainty and political polarization, drawing in those disillusioned with the mainstream and establishment alike. As with other radical factions, the danger lies not just in their rhetoric, but in their willingness to act on their beliefs, challenging the very foundations of law and order in the name of personal sovereignty.[292]

An example of the danger of this movement is the Bundy Standoff. In 2014 Cliven Bundy summoned a mob to his Nevada ranch and staged an armed

---

[290] Bruce Hoffman, *Inside Terrorism* (New York: Columbia University Press, 2006), 237.

[291] Institute for Strategic Dialogue, "Sovereign Citizens," *ISD Explainers*, October 3, 2022, accessed July 12, 2025, https://www.isdglobal.org/explainers/sovereign-citizens/.; Southern Poverty Law Center, "Sovereign Citizens Movement," *Extremist Files*, accessed July 12, 2025, https://www.splcenter.org/resources/extremist-files/sovereign-citizens-movement/.

[292] Federal Bureau of Investigation, "Domestic Terrorism: The Sovereign Citizen Movement," April 13, 2010, accessed July 12, 2025, https://archives.fbi.gov/archives/news/stories/2010/april/sovereigncitizens_041310/domestic-terrorism-the-sovereign-citizen-movement.

standoff over control of federal public land.[293] Although the standoff was resolved without bloodshed that has not always been the case.

Two pivotal events served as radicalizing moments:

- Ruby Ridge (1992): When federal agents killed Vicki Weaver during a siege at the family's cabin, she became a martyr in right-wing mythology.
- Waco (1993): The botched federal raid on the Branch Davidian compound, resulting in the deaths of 76 people including 25 children, became a rallying cry for anti-government extremists. Timothy McVeigh later cited Waco as the motivation for the 1995 Oklahoma City bombing.

Additionally, the apocalyptic novel *The Turner Diaries* deeply influenced the White Power imagination. It depicts a future race war and has served as a blueprint for revolutionary violence.

Like the extreme left, the White Power movement has not been able to move from the fringe to the mainstream. Instead, it has attempted to leverage real issues such as 2nd amendment rights, the prolife movement, and the so-called culture wars.

THE VANGUARD: From Compounds to Militias

The radical right has historically lacked the organizational discipline of its far-left counterparts. A few individuals—like Bob Matthews and Richard Butler—sought to create true revolutionary organizations, their efforts were routinely undermined by law enforcement. In the 1970s and 1980s, White Power militants began building rural compounds in preparation for social collapse. These static locations, however, made it easier for law enforcement to monitor and dismantle operations. Agencies had gained significant counterterrorism experience combating Marxist groups like the Weather Underground and M19CO, making the infiltration and dismantling of White Power cells relatively efficient.[294]

---

[293] Kirk Siegler, "A Decade after Armed Standoff, the Bundys Appear to Be Above the Law," *NPR*, June 20, 2024, accessed July 12, 2025, https://www.npr.org/2024/06/20/nx-s1-4966725/a-decade-after-armed-standoff-the-bundys-appear-to-be-above-the-law.; Southern Poverty Law Center, "War in the West: The Bundy Ranch Standoff and the American Radical Right," special report, July 10, 2014, accessed July 12, 2025, https://www.splcenter.org/resources/reports/war-west-bundy-ranch-standoff-and-american-radical-right.

[294] Smith, Brent, Terrorism in America: Pipe bombs and Pipe Dreams, State University of New York Press Albany 1994, 73.

Only one group, *The Order*, led by Bob Matthews, achieved sustained clandestine activity. Ultimately, Matthews died in a fiery standoff with federal agents on Whidbey Island, Washington. After that, White Power organizing was effectively suppressed for the remainder of the 1980s.

Militias and McVeigh

The 1990s witnessed a revival of right-wing activism through the militia movement. While much of it was rooted in survivalist culture, some groups became significantly more dangerous. The deadliest act of right-wing domestic terrorism was the 1995 Oklahoma City bombing, carried out by Timothy McVeigh and Terry Nichols. McVeigh, a Gulf War veteran, claimed no official ties to any movement. However, it is unlikely that he and Nichols acted entirely alone. There is evidence of McVeigh's connection to the Michigan Militia.[295]

Despite this, FBI Director Louis Freeh testified in 1997 that most militia groups were nonviolent.

Most of the militia organizations around the country are not, in our view, threatening or dangerous," FBI Director Louis J. Freeh testified before the Senate Judiciary Committee two days after a Denver jury convicted McVeigh of the bombing. The majority of militia members are nonviolent, and some have assisted the bureau in its investigations, he said.[296]

Compounds and Collapse

The *Aryan Nations* compound in Hayden Lake, Idaho, functioned as a hub for groups like the *Covenant, Sword, and Arm of the Lord* (CSA). Both were dismantled through federal prosecution. By the late 1990s, the militia movement again faded, largely due to infiltration by federal agents.

Although concern about right-wing extremism has risen sharply in the twenty-first century, distinguishing genuine threats from plots shaped or facilitated by FBI sting operations is often difficult. The 2020 conspiracy to kidnap Michigan Governor Gretchen Whitmer, for example, involved multiple undercover agents and informants, prompting continuing debate over the extent of the FBI's role and whether elements of the case approached entrapment. The mixed verdicts only deepened the controversy, allowing the case to be

---

[295] Aaron Brilbeck, "Michigan Militia Group Had Ties to Timothy McVeigh," *News 9*, October 9, 2020.

[296] Richard Leiby, "Many Militia Groups Scale Back, Distance Themselves From McVeigh," *Washington Post*, June 14, 1997, A08, https://www.washingtonpost.com/wp-srv/national/longterm/oklahoma/stories/militia.htm.

appropriated by radicals on both sides for their own political purposes.

Infiltration, Paranoia, and Radicalization

The three organizations most frequently cited as instigators of the January 6th Capitol Riot, the Proud Boys, Oath Keepers, and Three Percenters, were all already infiltrated by federal authorities.[297] Many of their leaders were known informants. This raises troubling questions about where authentic radicalism ends, and government provocation begins.

As in the 1970s, excessive federal action risks inflaming the very movements it seeks to suppress. Heavy-handed raids, pretrial detentions, and public demonization of nonviolent actors may undermine public confidence and strengthen the perception that the system is irreparably broken.

MASS UNREST AND CIVIL WAR FANTASIES

Most mass violence (as opposed to vanguard violence) by the radical right has taken the form of counter-protests or deliberate provocations rather than organic mass movements. Groups like the Proud Boys, Patriot Prayer, and the Oath Keepers claim to champion traditional American values, but showing up at political rallies with body armor and batons suggests otherwise. If not actively working in tandem with the White Power movement, these groups are certainly playing into their hands.

Efforts to recruit from broader society, such as the pro-life or Second Amendment movements, have largely failed to translate into mass mobilization. While prison gangs like the Aryan Brotherhood offer recruitment channels, these groups function more as criminal enterprises than political movements.

Eric Rudolph

The case of Eric Rudolph is often cited as proof of radical pro-life violence. However, Rudolph's actions—bombing an abortion clinic and the Atlanta Olympics—stemmed from a deeply personal ideology that mixed anti-government sentiment with religious extremism. As the FBI later concluded, he was a loner with no substantial support network.

Did anybody help Rudolph avoid detection?

Mr. Swecker: That's what a lot of people think. But Rudolph is such a loner that we strongly believed he simply wouldn't have trusted anybody. He had access to news; he had newspaper articles in his camp. He knew he was being pursued. I don't think

---

[297] Julie Kelly, *January 6: How Democrats Used the Capitol Protest to Launch a War on Terror Against the Political Right* (New York: Bombardier Books, 2022), Kindle, chap. 8.

he would have made himself vulnerable to being compromised or betrayed by letting anyone know where he was… He had borrowed ideas from a lot of different places and formed his own personal ideology. He clearly was anti-government and anti-abortion, anti-gay, 'anti' a lot of things. The bombings really sprang from his own unique biases and prejudices. He had his own way of looking at the world and didn't get along with a lot of people.[298]

Failed Public Mobilization: Charlottesville and Beyond

The 2017 Unite the Right rally in Charlottesville was the most significant public demonstration by White Power activists in recent memory. It was a public relations disaster. Violence, legal defeat, and widespread condemnation followed. The event's organizers were later bankrupted by a $25 million civil lawsuit. NPR reporter Ian Shapira concluded, "The white nationalist movement… has largely cratered."[299]

By 2021, attempts to stage "White Lives Matter" marches failed spectacularly, with many events cancelled or dwarfed by counterprotesters. [300] In some cases, antifa hoaxes were used to lure and expose white nationalists, further demoralizing the movement.

COUP D'ÉTAT: Potential, Fantasy, and Failure

Within the broader White Power and radical-right ecosystem, the idea of a coup d'état has persisted as a seductive fantasy, one rooted more in ideological storytelling than strategic plausibility. The coup d'état model, common in 20th-century revolutionary ideology, imagines that a relatively small number of well-placed actors can seize control of the state apparatus and reorient its power in line with their values.

However, a Coup Remains Unlikely. Despite these fantasies, the American far right has no access to institutional power on the scale necessary for a successful coup:

- The military and law enforcement remain largely professionalized and

[298] Eric Rudolph, "Eric Rudolph," *Federal Bureau of Investigation*, accessed November 20, 2025, https://www.fbi.gov/history/famous-cases/eric-rudolph.

[299] Ian Shapira, "What the 'Unite the Right' Trial Reveals about White Nationalism in the U.S.," *PBS NewsHour*, November 23, 2021, https://www.pbs.org/newshour/show/what-the-unite-the-right-trial-reveals-about-white-nationalism-in-the-u-s#audio.

[300] Ewan Palmer, "White Lives Matter Marchers Despondent After Failure: 'I Was the Only Person To Show Up,'" *Newsweek*, April 12, 2021, https://www.newsweek.com/white-lives-matter-marches-fail-protests-1582804.

constitutionalist, with mechanisms in place to root out radicalism.

- White Power groups are deeply fragmented, with competing theological, strategic, and ideological divisions (e.g., Christian Identity vs. Norse Paganism, separatists vs. constitutionalists).
- Their operational capacity is amateur at best, often drawing from militia or prepper cultures rather than disciplined revolutionary cadre networks.

Furthermore, the few instances where radicals have attempted to infiltrate or influence state institutions, whether via military enlistment or political office, have not yielded sufficient leverage to even initiate, let alone execute, a coup.

The Appeal of the Coup Myth

Still, the symbolic appeal of a coup endures for several reasons:

- It offers a shortcut to power for those disillusioned with electoral politics.
- It reinforces an apocalyptic worldview that imagines the U.S. is already in terminal decline.
- It builds upon the belief that a single destabilizing event could tip society into chaos.

Some adherents reinterpret events like the January 6 Capitol riot through this lens, despite its clear lack of organization or command structure. What was, in reality, a disorganized protest-turned-mob, is framed by extremists as a "test run" or "dry rehearsal" for a larger, future insurrection. If the riot was orchestrated by the radical right, it was a total failure as a coup and a strategic disaster, resulting in arrests, organizational collapses, and loss of public sympathy.

REVOLUTIONARY VICTORY AND CONSOLIDATION OF POWER

Whether orchestrated or spontaneous, such crises as the January 6 riot can be used to justify heavy-handed government crackdowns. Yet, history warns that overzealous responses can deepen divisions rather than foster national unity. In times of turmoil, mercy and restraint do more to undermine radicalism than vengeance ever could.

As for the prospect of the White Power movement seizing power, the scenario is both grim and highly improbable. Should such a movement ever succeed, the result would be catastrophic, marked by violence and chaos. Fortunately, the movement lacks the numbers, institutional foothold, and organizational discipline required to capture and hold power. While their rhetoric of a so-called "Racial Holy War" (RaHoWa) remains largely confined to the fringes, their capacity for violence, though limited, still poses a real and disproportionate threat. Their dystopian ambitions are unlikely to materialize, but their potential for disruption should not be underestimated.

TYRANNY

A radical-right utopia regime would likely resemble their cult-like compounds, paranoid, ideologically incoherent, and authoritarian. The Branch Davidian and the Aryan Nations compounds offer glimpses into what governance under such a regime would look like: reclusive, brutal, and short-lived.

EXPORT AND EXPANSION

Fears of international White Supremacy are largely overstated. Aside from occasional online contact, the American White Power movement has little global coordination. Notably, one prominent European neo-Nazi network was run by a 13-year-old Estonian boy.[301] By contrast, far-left anarchist groups enjoy substantial international infrastructure. The real danger, however, lies in reciprocal radicalization, where violence on one side fuels extremism on the other, and in ideological convergence.

Accelerationism and the Boogaloo Movement

The Boogaloo movement exemplifies the concept of ideological convergence.[302] Adherents are often armed, anti-government, and engaged in both left- and right-wing protests. The Southern Poverty Law Center and Brookings both classify Boogaloo as accelerationist, those who seek to ignite civil war to collapse the existing order. The Brookings Institute cite James Mason, a neo-Nazi who advocated chaos as a strategy to usher in fascism. His *Siege* newsletter remains influential in these circles.[303]

Boogaloo adherents, however, claim they are not White Supremacist, and

---

[301] "Far-Right Terrorist Ringleader Found to Be Teenager in Estonia," *Deutsche Welle*, April 22, 2020, https://www.dw.com/en/far-right-terrorist-ringleader-found-to-be-teenager-in-estonia/a-53085442.

[302] Creede Newton, "The Boogaloo Movement Has a New Strategy," *Al Jazeera*, April 16, 2021, https://www.aljazeera.com/news/2021/4/16/the-boogaloo-movement-has-a-new-strategy.

[303] Amy Beeman, "The Boogaloo Movement: 5 Fast Facts You Need to Know," *Heavy*, last updated June 4, 2020, https://heavy.com/news/2020/06/the-boogaloo-movement-5-fast-facts/.

have joined BLM in anti-police protests.

Boogbois have often walked a fine line, attempting to position themselves as an armed buffer between police and demonstrators, peacekeepers between leftist and rightist groups, as well as protectors of private property...as principled as some of its members may sound, the Boogaloo meme at the center of the movement is still one of impending war. [304]

It may be less important to label the Boogaloo movement as leftist or rightist than to understand the threat posed those by who are actively seeking to incite a civil war in the United States.

Summary of the Radical Right

As Robert Gerwarth observed on Benito Mussolini's rise, Violence can trump democracy.[305] Power, once unleashed, transcends ideology. Hitler found the fear communists could propel him to power.[306] The chaos of the Russian Revolution created considerable and not misplaced anxiety in western Europe. The fascists offered stability and law and order while the German Communists offered a Bolshevik revolution. With the success of the Nazis and the purge of the communists following the Reichstag fire many communist street fighters changed sides and joined the Nazis. Sepp Dietrich, who would later command the SS Leibstandarte division, was himself once the elected leader of a communist soldiers' council in 1918.[307]

America, too, has seen echoes of this dynamic. The aftermath of Vietnam, much like post-WWI Germany, left many veterans searching for meaning and susceptible to radical ideologies. Today, veterans of Iraq and Afghanistan can be found on both the far left and right, a phenomenon Eric Hoffer warned about: disillusioned soldiers are prime recruits for mass movements.[308] If government institutions fail to address genuine economic, cultural, or institutional grievances, those frustrations become fertile ground for radicals. When the center cannot hold, it is the extremists who rush in to fill the void.

There is no shortage of anxiety about the specter of right-wing domestic

---

[304] Zach Weissmueller, "The Complicated Truth About the Boogaloo Movement," *Reason*, October 16, 2020, video, https://reason.com/video/2020/10/16/the-complicated-truth-about-the-boogaloo-movement/.

[305] Robert Gerwarth, *The Vanquished: Why the First World War Failed to End, 1917-1923*, (London: Penguin books, 2017), 164.

[306] Ibid, 153-167.

[307] Ibid, 217.

[308] L, Eric. *The True Believer: Thoughts on the Nature of Mass Movements* (New York: Harper & Row, 1966).

terrorism in the 21st century, but separating paranoia from reality is no easy task. Groups like Atomwaffen Division do pose a threat, but they remain tiny and largely isolated. Efforts to link the broader White Power movement to violent plots have often been muddied by government involvement; the 2020 plot to kidnap Michigan Governor Gretchen Whitmer, for instance, was revealed to be an FBI sting, planned, funded, and largely orchestrated by federal agents themselves.[309]

Heavy-handed government responses risk fueling the very radicalism they seek to suppress. When law enforcement employs military-style raids against citizens accused of insurrection, as seen in the aftermath of January 6, it can erode public trust and feed the narrative that the system is rigged. The extended pretrial incarceration of some January 6 defendants raised further questions about due process. History shows that the government is most effective at countering extremism when it adheres strictly to the rule of law and constitutional principles. Overreach, whether through COINTELPRO, Waco, or Ruby Ridge, has only served to create martyrs and deepen polarization on both the left and right. Ultimately, legitimacy is best preserved not through force, but through justice and restraint.

RTVC Table: American Radical Right

RTVC Phase	American Far Right Movement
Ideological Development	Neo-Nazism, Christian Identity, Anti-Federalism, Accelerationism
The Vanguard	The Order, Aryan Nations, CSA
Mass Civil Unrest	Unite the Right (2017), January 6 (2021), Militia rallies
Civil War (Optional)	Promoted by Boogaloo and Accelerationist factions,

309 Kelly, *January 6*, 265.; Branko Marcetic, "We Shouldn't Trust the FBI's Narrative on the Gretchen Whitmer Kidnapping Scheme," *Jacobin*, October 20, 2020, https://www.jacobinmag.com/2020/10/gretchen-whitmer-kidnapping-fbi-domestic-terrorism.; John Agar, "FBI, Informants Drove Plot to Kidnap Gov. Whitmer, Charges Should Be Dropped, Defendants Say," *MLive*, December 2021, https://www.mlive.com/news/grand-rapids/2021/12/fbi-informants-drove-plot-to-kidnap-gov-whitmer-charges-should-be-dropped-defendants-say.html.; Chuck Goudie, Barb Markoff, Christine Tressel, and Ross Weidner, "Michigan Governor Kidnap Plot Dreamt Up by FBI Informants, Accused Leader Claims," *ABC 7 Chicago*, August 13, 2021, https://abc7chicago.com/mi-governor-kidnap-plot-michigan-gov-gretchen-whitmer/10948197/.

RTVC Phase	American Far Right Movement
	but not materialized
Coup d'État (Optional)	Lacks institutional access or command structure; fantasy outweighs feasibility
Revolutionary Victory	Hypothetical only; lacks necessary mass support
Tyranny	Envisioned theocratic or pagan authoritarian enclaves
Export & Expansion	Mostly online; weak international connections; risk of ideological convergence with leftist radicals
Democratic Transition	Not applicable; movement seeks to replace, not reform, the system

CONCLUSION

The last really great bourgeois-haters died at about the same time: Sartre, De Gaulle, and Heidegger. (Americans are not sufficiently aware that hatred of the bourgeois is at least as much a thing of the Right as of the Left.)[310]

Revolution may seem exciting to the young, and idealistic, but the truth is these are bloody affairs resulting in almost unimaginable misery. While it may be a colloquialism to say one man's terrorist is another's freedom fighter, those seeking freedom and justice in the United States do not and should not resort to violence. Those who advocate violence are not seeking liberty; they are seeking power. And those who succumb to the siren call of terror rarely, if ever, achieve a truly free society. As Theda Skocpol observed, "'Radical' connotes for all too many Americans mindless, even violent, militancy." Yet, the leaders of radical movements are rarely mindless; there is a calculated logic to their use of violence.

However, just the cause, the authority to take up political violence in a republic belongs ultimately to the sovereign people, not to those who have appointed themselves its executors. The Weathermen, like radicals on both the far left and right, acted as vigilantes, placing themselves above the law and the

---

[310] Allan Bloom, *The Closing of the American Mind: How Higher Education Has Failed Democracy and Impoverished the Souls of Today's Students* (New York: Simon & Schuster, 1987), 159.

collective will. Cicero reminds us that the foundation of good government is popular consent and adherence to the rule of law.[311] When individuals or groups become a law unto themselves, society descends into chaos, everyone doing "what is right in his own eyes."[312] The rule of law is a public trust, a safeguard against the tyranny of private judgment and unchecked bias.

Today, as confidence in public judgment and institutional effectiveness wanes, more citizens are tempted to take matters into their own hands whether through civil disobedience, anti-lockdown protests, or anti-deportment riots. When political judgment collapses, public space fragments into countless private crusades, leaving only vigilante justice.[313]

In *Uncivil Agreement*, Lilliana Mason advances the argument that contemporary American political polarization is driven less by widening policy disagreement than by the convergence of partisan affiliation with powerful social identities such as race, religion, culture, and geography.[314] This process of "social sorting" has transformed political parties into markers of group belonging, intensifying affective polarization even as issue positions remain relatively stable. Politics, Mason contends, now functions as an arena of group competition in which partisan cues signal "who people are," rendering opposing parties socially distant and emotionally threatening rather than merely ideologically different.

Consequently, political engagement is increasingly shaped by emotion, particularly anger and resentment, rather than reasoned evaluation of policy. Mason shows that this identity-based conflict normalizes incivility, weakens support for democratic norms, and reduces incentives for compromise, which comes to be viewed as disloyalty to one's group. In this framework, ethical breaches and norm violations are often excused when committed by in-group actors, deepening asymmetries in democratic accountability. For comparative analyses, Mason's work is particularly valuable in reframing polarization as a structural and psychological phenomenon rooted in social identity, highlighting why institutional reforms or informational corrections alone are insufficient to

---

311 Jill Harries, "Servius, Cicero and the Res Publica of Justinian," in *Cicero's Law: Rethinking Roman Law of the Late Republic*, ed. Paul J. du Plessis (Edinburgh: Edinburgh University Press, 2016), 134.

312 *Judges 17:6* (English Standard Version).

313 Brad Littlejohn, "The Dark Knight and the Possibility of Political Judgment," *Political Theology Network*, July 31, 2012, 1. https://politicaltheology.com/dark-knight-possibility-political-judgment/.

314 Lilliana Mason, *Uncivil Agreement: How Politics Became Our Identity* (Chicago: University of Chicago Press, 2018).

temper political conflict in highly sorted societies.

Disturbingly, calls to overthrow the U.S. government now echo from both the left and right. What was once the "fringe of the fringe" is inching toward the mainstream, fueled by divisive politics, economic stress, and the evolving playbook of radical activists. The radical left and right share more than they admit: both seek to hijack mainstream politics for their own revolutionary ends. While the right tends to be more ethnically homogenous and the left more diverse, both rely on exploiting grievances and inflating their influence through alliances with broader protest movements. The left has historically been more disciplined, better organized, and more successful at evading law enforcement, sometimes aided by foreign support. The right, meanwhile, has struggled to build mass movements, creating front groups or resorting to infiltration of legitimate protests, as seen in Charlottesville.

January 6 resists easy classification. A coup d'état requires the seizure of state power by insiders with institutional leverage, a capacity the participants did not possess. Whether the event meets the legal threshold for insurrection remains contested in both courts and scholarship. Nor is it clear who was responsible for the chaos that day. What is clear is that it exhibited the defining features of the RTVC's mass mobilization stage: ideological motivation and coordination by organized cells as well as the overreaction by the security forces which exacerbated the situation. Its significance for this study lies not in its legal label but in what it reveals about the fragility of constitutional order when reciprocal radicalization reaches critical mass.

The distinction between protest and insurrectionary mobilization is often revealed through material preparation. When demonstrators arrive equipped for confrontation rather than expression, they signal a tactical intent that moves beyond mass civil unrest and toward organized revolutionary or counterrevolutionary action. At the same time, government overreach, whether in COINTELPRO, Waco, or the militarized response to January 6, often creates martyrs and deepens political polarization. The legitimacy of a republic depends upon fidelity to the rule of law, not the heavy hand of state power.

The American constitutional republic, with its checks and balances, is designed to keep both governmental and popular power in check. When that balance is lost, when power is concentrated in the hands of the few, whether radicals or bureaucrats, the republic ceases to serve its true sovereign: the people. Popular sovereignty is sustained not by categories or factions, but by neighbors trusting neighbors, and by civic engagement at the local level. It is the moderate, engaged citizen, the true middle class, who will ensure our nation's survival. Move too far left or right, and you arrive at the same destination: tyranny. The desire for a better world is nearly universal; it is the resort to destruction that divides revolutionaries from the people they presume to lead.[315]

Both wings of radicalism have weaponized social media, exploiting algorithms that amplify outrage and deepen polarization. Doxxing, deplatforming, and conspiracy echo chambers drive the cycle of reciprocal radicalization. As centrists and moderates find themselves politically homeless, the risk of political violence and illiberalism grows. If polarization continues unchecked, and institutional trust erodes, the American Republic may be imperiled

Reciprocal radicalism is no accident, it is a process of mutual escalation, fueled by weak institutions, irresponsible media, and ideological absolutism. The lessons of Weimar and 1960s America should not just serve as warnings, but as guides for breaking the cycle. In the end, it is only through renewed trust, civic engagement, and steadfast commitment to the rule of law that we can preserve the republic and avoid the twin perils of radicalism and repression.

There is also a disturbing confluence of the extreme left and right, both denounce Zionists who they believe control the U.S. government. Since this chapter was first written the radical right has made a resurgence evident in the so-called "Groyper" movement and influencers Nick Fuentes. With the inexcusable help of figures such as Tucker Carlson and Candace Owens, right-wing antisemitism has come out of the shadows and given the radical left the enemy it has imagined and hoped for. And yet the radical left has taken up the cause of Palestine and developed violent antisemitism. Both extremes now imagine the same enemy, Z.O.G.

---

[315] Lennon, John. "Revolution." Recorded 1968. Apple Records 5099969942877, 1968, 7-inch single.

CONCLUSION

BEYOND LEFT AND RIGHT

Conventional wisdom holds that American politics is neatly split between Conservative Republicans on the Right and Liberal Progressive Democrats on the Left. In reality, both the Republican and Democrat parties are sprawling coalitions, each uniting a wide array of sometimes conflicting interests in order to amass enough influence to advance their agendas. These agendas are not monolithic; rather, they are composites, shaped by the diverse priorities of the groups that comprise each party. This explains why seemingly unrelated issues such as abortion, gun rights, hate speech, economic policy, border security, education (charter vs. public schools), evolution, climate change, religion, and foreign policy, become bundled together under a single partisan banner.

The familiar labels, "progressive" and "conservative," are ultimately misleading. The true fault line in American politics is not between liberal and conservative, but between **centralization and decentralization of power**. The opposite of "conservative" is not "progressive," but "radical." Nor is "conservative" synonymous with "reactionary." Reactionaries long to return to a bygone era; conservatives prefer to move forward with caution and deliberation, while radicals are eager to press ahead without restraint.

The distinction between liberals and conservatives is often a matter of degree, but the deeper question concerns authority: where should power reside? While we may agree in principle that sovereignty rests with the people, do citizens actually possess the means to exercise that sovereignty? Excessive centralization tends to erode popular sovereignty, while excessive decentralization risks chaos and lawlessness. The enduring debate is over how much power should be vested in the federal government, how much reserved for the states, and what decisions should be left to local communities.

This centralizer-versus-decentralizer division exists within each of the two parties. Among Republicans, the centralizing faction is represented by "establishment Republicans" and the so-called Neo-Conservatives or, pejoratively, RINOs (Republicans In Name Only). Traditional conservatives frequently oppose the Neo-Cons as vigorously as they do the Democrats. A third group within the party, the Populists, may support either centralization or decentralization, but their allegiance is typically driven more by emotion and cultural identity than by consistent ideology.

The Democrat Party is similarly divided. Establishment or moderate progressives seek to control the system, not overturn it; radical progressives, by contrast, wish to dismantle existing structures and start anew. Both factions, however, tend to favor centralization, differing primarily in the pace and scope of consolidation they pursue. A third group, the moderate liberal Democrats, are often closer in outlook to conservative Republicans than to either of the other Democrat factions, particularly on questions of institutional process and constitutional limits.

The Neo-conservative support for centralized governance and interventionist foreign policy places them closer to moderate progressives on the centralization axis than to traditional conservatives who emphasize constitutional restraint. This shared outlook explains why "Never Trump" Republicans found common cause with establishment Democrats during the first Trump administration. Radical progressives, though outside the establishment, pursue the same goal of centralized control, albeit through more extreme means, revolutionary transformation rather than incremental reform. All parties profess a desire for a just society, but they differ sharply on how to achieve it: through centralization or decentralization, through gradual reform or radical transformation, and, crucially, through their understanding of human nature.

The Centralization Imperative and Revolutionary Logic

The centralization-decentralization dynamic shapes revolutionary outcomes. The RTVC framework's tyranny stage typically involves power consolidation, whether in Lenin's Politburo, Khomeini's Supreme Leadership, Robespierre's Committee of Public Safety, or Cromwell's Protectorate. Revolutionary movements gravitate toward totalitarianism because their ideology tells them their goal is so important that it justifies any means necessary. Therefore, they consolidate power through centralization, eliminating rivals and foreclosing the democratic pluralism they claimed to champion.

Chapter 4's analysis of reciprocal radicalism in contemporary America reveals this pattern in microcosm. Both the Radical Left (Weather Underground, Antifa) and Radical Right (white nationalist militias, Patriot Front) pursue the same end despite opposing ideologies: centralized control of the state to impose their vision on society. The Radical Left seeks centralization through revolutionary transformation of existing institutions, using state power to enforce equity mandates, redistribute wealth, and suppress dissenting speech as "hate." The Radical Right seeks centralization using state power to enforce cultural uniformity and suppress dissenting movements as "un-American."

Neither faction accepts the constitutional framework of dispersed power and mediating institutions. Both view their opponents as existential threats rather than fellow citizens with whom they disagree. Both increasingly resort to extra-constitutional means, street violence, intimidation, doxing, deplatforming, when institutional processes produce unacceptable outcomes. Both delegitimize elections and judicial decisions that go against them, framing defeats as evidence of systemic corruption rather than democratic setbacks requiring renewed persuasion.

The centralization/decentralization tension illuminates structural tendencies across diverse political contexts, but it should not be mistaken for a universal explanatory key. A European Union regulatory dispute, a jihadist insurgency, and a domestic protest movement operate at different scales, involve different actors, and respond to different incentive structures. What the axis reveals is a shared "directionality": movements and institutions that concentrate decision-making authority in unaccountable elites tend toward coercion, while those that distribute authority among competing institutions tend toward stability. The axis identifies this tendency without claiming that all forms of centralization are equivalent or that the dynamics producing centralization in Brussels and in Tehran are the same.

The RTVC applies most clearly to utopian revolutions, movements that claim to inaugurate an ideal political order. Volume 2 will examine, among other things, revolutions that did not rely on revolutionary terror. The Solidarity movement in Poland and the Velvet Revolution in Czechoslovakia are two notable examples. Both demonstrate how disciplined nonviolent movements can produce successful democratic transitions.

The moderate center, traditional conservatives who favor gradual reform through constitutional processes and moderate liberals who support deliberative democracy, finds itself under assault from both extremes. As Chapter 4 demonstrates through the Weimar parallel, this pattern of competing centralizers destroying the constitutional middle ground is not unique to contemporary America. It is a recurring feature of revolutionary situations throughout history.

When mediating institutions such as churches, civic organizations, labor unions, local governments, and responsible journalism, weaken then citizens face only the binary choice between radical transformation and authoritarian reaction. Constitutional politics becomes impossible. The American Founders understood this danger, which is why they designed a system that disperses power and protects minority rights even against democratic majorities. Federalism, separation of powers, bicameralism, staggered terms for politicians, and supermajority requirements for constitutional amendments, all serve to prevent any faction from rapidly centralizing power.

Yet these constitutional safeguards depend on a culture of forbearance and mutual toleration. When political opponents become enemies to be crushed rather than fellow citizens to be persuaded, constitutional procedures appear as obstacles rather than as protection. The centralizing impulse whether justified by revolutionary justice, historical progress, national security, or public health overwhelms institutional resistance. To paraphrase Friedrich Hayek, the road to serfdom is paved with good intentions; centralization proceeds through emergency measures that never expire, temporary powers that become permanent, and exceptional cases that become routine.

Revolutions and Constitutions

The American Revolution succeeded constitutionally because the Founders explicitly designed a system dispersing power across federal and state governments, three branches, and two legislative chambers, and enshrining individual rights. The French Revolution descended into Terror because revolutionaries concentrated all power in the National Convention, then in the Committee of Public Safety, eliminating mediating institutions between the revolutionary state and individual citizens. The English Civil War demonstrated both trajectories: Cromwell's centralized Protectorate produced tyranny, but the Glorious Revolution's constitutional settlement dispersed power through parliamentary checks on monarchy.

Contemporary American polarization reflects the failure to maintain this constitutional balance. Both parties increasingly pursue centralization when they control power by executive orders circumventing Congress, judicial activism imposing national standards, and administrative agencies promulgating regulations without legislative authorization. The centralizing ratchet operates regardless of which party controls Washington. Each expansion of federal authority becomes the baseline for the next administration, regardless of partisan rhetoric about limited government or states' rights.

The mediating institutions that once moderated this centralizing tendency have weakened precisely when their moderating influence is most needed. Churches that once commanded moral authority and provided social services independent of the state have secularized or politicized, becoming either culturally irrelevant or partisan activists. Labor unions that once gave workers voice in economic decisions have declined, or been captured by radicals, reducing countervailing power against both corporate and state centralization. Local newspapers that held municipal governments accountable have disappeared, reducing transparency. Civic organizations that brought diverse citizens together in common cause, Rotary clubs, Elks lodges, bowling leagues, have atrophied, reducing social cohesion.

This institutional erosion creates the preconditions for revolutionary mobilization. When citizens lack non-state institutions through which they may exercise agency, the state becomes the sole arena for political activity. When the state becomes the sole arena, politics becomes total war. When politics becomes total war, constitutional restraints appear as impediments rather than protections. And when constitutional restraints fail, the RTVC pattern recurs: ideological development (us vs. them narratives), vanguard formation (radical factions prepared for violence), mass mobilization (protest movements), revolutionary victory (seizure of state institutions), consolidation (elimination of rivals), and tyranny (permanent emergency rule).

America has not completed this progression to revolution, but warning signs are unmistakable. The question is whether constitutional culture and institutional resilience remain strong enough to reverse the trajectory or whether reciprocal radicalization will accelerate until mediating options disappear.

External Interference and the Sovereignty Dilemma

Volume I has established that revolutions are contests over sovereignty, who holds power, how it is legitimated, whether it will be centralized or dispersed. These contests unfold primarily within states, driven by internal dynamics of grievance, mobilization, and institutional failure. Yet few revolutions occur in isolation. External powers intervene, either covertly or overtly, to shape outcomes, pursue strategic interests, or advance ideological agendas.

Volume II examines this external dimension. How do foreign powers navigate the tension between supporting local movements and advancing their own interests? When does external support enable indigenous success, and when does it undermine local legitimacy? Under what circumstances, if any, is intervention justified and how should it be conducted to avoid repeating the pattern of well-intentioned interventions producing disastrous outcomes?

SELECTED BIBLIOGRAPHY

CLASSICAL AND ANCIENT SOURCES

Aristotle. *Politics.* Translated by Benjamin Jowett. Oxford: Clarendon Press, 1885.

Cicero, Marcus Tullius. *De Re Publica* (On the Republic). Translated by Clinton W. Keyes. Loeb Classical Library. Cambridge, MA: Harvard University Press, 1928.

Dionysius of Halicarnassus. *Roman Antiquities.* Translated by Earnest Cary. Loeb Classical Library. Cambridge, MA: Harvard University Press, 1937.

Plutarch. *Lives.* Translated by John Dryden. Revised by Arthur Hugh Clough. New York: Modern Library, 2001.

Polybius. *The Histories.* Translated by Robin Waterfield. Oxford: Oxford University Press, 2010.

Suetonius. *Lives of the Caesars.* Translated by Catharine Edwards. Oxford: Oxford University Press, 2000.

Tacitus, Cornelius. *The Histories* and *The Annals.* Translated by Clifford H. Moore and John Jackson. Loeb Classical Library. Cambridge, MA: Harvard University Press, 1925-1937.

ENLIGHTENMENT AND EARLY MODERN POLITICAL THOUGHT

Burke, Edmund. *Reflections on the Revolution in France.* Edited by J.G.A. Pocock. Indianapolis: Hackett Publishing, 1987.

Grotius, Hugo. *The Rights of War and Peace.* Edited by Richard Tuck. Indianapolis: Liberty Fund, 2005.

Hobbes, Thomas. *Leviathan.* Edited by Richard Tuck. Cambridge: Cambridge University Press, 1996.

Hume, David. *Essays: Moral, Political, and Literary.* Edited by Eugene F. Miller. Indianapolis: Liberty Fund, 1987.

Locke, John. *Two Treatises of Government.* Edited by Peter Laslett. Cambridge: Cambridge University Press, 1988.

Montesquieu, Charles de Secondat, Baron de. *The Spirit of the Laws.* Translated and edited by Anne M. Cohler, Basia Carolyn Miller, and Harold Samuel Stone. Cambridge: Cambridge University Press, 1989.

Paine, Thomas. *Common Sense and Other Writings.* Edited by Gordon S. Wood. New York: Modern Library, 2003.

Rousseau, Jean-Jacques. *The Social Contract and Other Later Political Writings.* Edited and translated by Victor Gourevitch. Cambridge: Cambridge University Press, 1997.

Smith, Adam. *The Wealth of Nations.* Edited by Edwin Cannan. New York: Modern Library, 1994.

AMERICAN FOUNDING DOCUMENTS AND WRITINGS

Hamilton, Alexander, James Madison, and John Jay. *The Federalist Papers.* Edited by Clinton Rossiter. Introduction by Charles R. Kesler. New York: Signet Classics, 2003.

Jefferson, Thomas. *The Declaration of Independence and Other Writings.* Edited by Joyce Appleby. New York: Penguin Books, 2012.

Madison, James. *Notes of Debates in the Federal Convention of 1787.* Introduction by Adrienne Koch. New York: W.W. Norton, 1987.

Washington, George. *George Washington: Writings.* Edited by John Rhodehamel. New York: Library of America, 1997.

Witherspoon, John. *Lectures on Moral Philosophy.* Edited by Jack Scott. Newark: University of Delaware Press, 1982.

MODERN SCHOLARSHIP - POLITICAL THEORY AND HISTORY

Belew, Kathleen. *Bring the War Home: The White Power Movement and Paramilitary America.* Cambridge, MA: Harvard University Press, 2018.

Gerwarth, Robert. *The Vanquished: Why the First World War Failed to End.* New York: Farrar, Straus and Giroux, 2016.

Glubb, John Bagot. *The Fate of Empires and Search for Survival.* Edinburgh: William Blackwood & Sons, 1978.

Goldstone, Jack. *Revolution and Rebellion in the Early Modern World.* Berkeley: University of California Press, 1991.

Hankins, James. "Exclusivist Republicanism and the Non-Monarchical Republic." *Political Theory* 38, no. 4 (2010): 452-482.

Hayek, Friedrich A. *The Road to Serfdom.* Chicago: University of Chicago Press, 1944.

Huntington, Samuel P. *The Clash of Civilizations and the Remaking of World Order.* New York: Simon & Schuster, 1996.

Mason, Lilliana. *Uncivil Agreement: How Politics Became Our Identity.* Chicago: University of Chicago Press, 2018.

Potter, David. *Emperors of Rome: The Story of Imperial Rome from Julius Caesar to the Last Emperor.* London: Quercus, 2007.

Robinson, Glenn. "Movements of Rage: Political Violence and Armed Groups." *Journal of Democracy* 25, no. 4 (2014): 23-39.

Schaub, Diana. "Captain Kirk and the Art of Rule." In *The Political Science Reviewer,* Vol. 42. New Brunswick, NJ: Transaction Publishers, 2018.

Skocpol, Theda. *States and Social Revolutions: A Comparative Analysis of France, Russia, and China.* Cambridge: Cambridge University Press, 1979.

Sowell, Thomas. *The Quest for Cosmic Justice.* New York: Free Press, 1999.

Spengler, Oswald. *The Decline of the West.* Translated by Charles Francis Atkinson. 2 vols. New York: Alfred A. Knopf, 1926-1928.

Tilly, Charles. *From Mobilization to Revolution.* Reading, MA: Addison-Wesley, 1978.

Toynbee, Arnold J. *A Study of History.* 12 vols. Oxford: Oxford University Press, 1934-1961.

MILITARY THEORY AND IRREGULAR WARFARE

Marighella, Carlos. *Minimanual of the Urban Guerrilla.* Havana: Tricontinental, 1970.

Myers, Thomas. "Terrorist to Tyrant." *Homeland Security Affairs* 7, art. 6 (February 2011).

Rubin, David. "Transforming Intelligence Using Industry Best Practices: The Terrorism Value Chain." *Homeland Security Affairs* 3, no. 1 (2007).

ADDITIONAL SOURCES

Al-Banna, Hasan. *Five Tracts of Hasan al-Banna* (1906-1949). Translated by Charles Wendell. Berkeley: University of California Press, 1978.

Ayers, Bill. *Fugitive Days: A Memoir.* Boston: Beacon Press, 2001.

Qutb, Sayyid. *Milestones.* Edited by A.B. al-Mehri. Birmingham, UK: Maktabah Booksellers and Publishers, 2006.

INDEX

ABOUT THE AUTHOR

Thomas Ripley Myers is an instructor of Professional Military Education. He is a retired soldier who served in the Croatian Defense Forces, the French Foreign Legion, and in the U.S. Army Infantry and Special Forces. He holds an AS in Military Science from Valley Forge Military College, a BS in History from Eastern University, an MS in Strategic Intelligence, and an MA in Ancient and Classical History from American Military University. He serves in his local church. Enjoys Skiing, Karate and spending time with his family.

ACKNOWLEDGEMENTS

This book was edited with the assistance of artificial intelligence. Following the completion of the manuscript, I used Claude for grammar and stylistic flow, and ChatGPT for a subsequent editorial review. I am grateful to Dr. James Lacey for making his Peer Review Helper available through that same platform.

Lawrence Turner read several draft versions with patience and care. He is always there for a brother.

The ideas developed in these pages have been tested and refined in countless conversations with colleagues I cannot adequately enumerate here. I owe particular debts to those who served alongside me in A/4/10 SFG(A) and to the faculty and students of the Enlisted Academy, whose practical wisdom sharpened arguments that might otherwise have remained purely academic.

Finally, Rebecca not only read the earliest drafts with a careful eye but gave me the sustained, undisturbed time that writing and revision demand. This book would not exist without her patience and support. It is dedicated to her for good reason.

All errors that remain are mine alone.

www.ingramcontent.com/pod-product-compliance
Lightning Source LLC
LaVergne TN
LVHW040515210926
848480LV00005BA/1465
* 9 7 9 8 9 9 4 0 4 1 2 2 2 *